Ion-Pair Chromatography and Related Techniques

ANALYTICAL CHEMISTRY SERIES

Series Editor

Charles H. Lochmüller
Duke University

ANALYTICAL CHEMISTRY SERIES

Ion-Pair Chromatography and Related Techniques

Teresa Cecchi

CRC Press
Taylor & Francis Group
Boca Raton London New York

CRC Press is an imprint of the
Taylor & Francis Group, an **informa** business

CRC Press
Taylor & Francis Group
6000 Broken Sound Parkway NW, Suite 300
Boca Raton, FL 33487-2742

First issued in paperback 2017

© 2010 by Taylor and Francis Group, LLC
CRC Press is an imprint of Taylor & Francis Group, an Informa business

No claim to original U.S. Government works

ISBN 13: 978-1-138-11206-3 (pbk)
ISBN 13: 978-1-4398-0096-6 (hbk)

Library of Congress Cataloging-in-Publication Data

Cecchi, Teresa.
   Ion-pair chromatography and related techniques / Teresa Cecchi.
      p. cm. -- (Analytical chemistry series)
   Includes bibliographical references and index.
   ISBN 978-1-4398-0096-6 (hardcover : alk. paper)
   1. Chromatographic analysis. I. Title. II. Series.

QD79.C4C35 2010
543'.8--dc22   2009031933

**Visit the Taylor & Francis Web site at
http://www.taylorandfrancis.com**

**and the CRC Press Web site at
http://www.crcpress.com**

Dedication

To Andrea, drop of fading eternity.
To Euridice and Gaia, life for my life.

Contents

Preface

Ion-Pair Chromatography (IPC) is an intriguing separation mode of analytical chemistry. IPC is not a passing fad, rather it is an advance that will have lasting impact since it has an honorable place in the HPLC armory for tackling difficult analyses of organic and inorganic ions, ionogenic, neutrals and zwitterionic compounds.

Despite its impact and progress, IPC has not been the focus of any recent books; this prompted the author to conceive and write this book. Distilling the knowledge gained from preeminent research, *Ion-Pair Chromatography and Related Techniques* deals with the basics of this established and easily tunable technique and with its rapid evolution due to the introduction of novel and challenging ion-pair reagents and strategies. It provides crucial tips, decisive skills, awareness, and advice to both experienced and novice chromatographers. Emphasis is placed on the progress from theoretical modeling of the IPC physico-chemical separation principle to application, to provide a clear understanding and an easy description of the multiplicity of interactions involved in an IPC system, and to help the chromatographer to perform educated guesses during method development.

This book is also aimed at presenting a broad outline of the recent scope of the application of this separative strategy and at establishing order in the complex welter of IPC separations. Many outstanding recent reports were selected for inclusion in this book to prove the substantial and practical potential of IPC in the life sciences, medicine, pharmacology, forensic, food, and environmental sectors, as well as to provide an up-to date overview of this chromatographic method.

The enduring upgrade of column technology and instrumentation to improve data quality and to increase sample throughput via ultra-fast separations is also described. The relation of IPC to other instrumental methods of analysis, the translation of the ion-pair concept in different analytical techniques, and non-separative aims of IPC, that are topics for which there is a dearth in the scientific literature, are critically discussed.

Future investigation needs and stimulating research suggestions are submitted to the separation scientists' community with the aim of advancing this exciting separative method and raising the level of interest in valuable and still unexplored analytical schemes.

The Author

Teresa Cecchi earned a PhD in chemistry from Camerino University in Italy in 1997. She focused on environmental chemistry at Institute Fresenius Gruppe, Germany and served as a consultant to food chemistry laboratories. After working as a researcher at Camerino University on the SUPREME project titled "Pigmentation in South American Camelids," she concentrated on ion-pair chromatography. Her major research interests encompass many aspects of this technique including retention modeling, unusual analytes such as zwitterions, and application of this technique to non-separative functions.

Dr. Cecchi's career and studies also span the fields of food packaging chemistry, natural dyes, electrochemistry, and the antioxidant activities of nutraceuticals. She acted as the organizer of a research group at Istituto Tecnico Industriale Statale (ITIS) Montani in Italy and taught as a contract professor at La Sapienza University in Rome. She is currently a contract professor on the science and technology faculty at Camerino University, teaching master's level courses in chemistry and advanced chemical methodologies. She is also a professor of analytical chemistry at ITIS Montani.

Dr. Cecchi is also involved in "Teaching of Experimental Sciences" and "Scientific Master Degrees," two projects whose purposes are, respectively, to improve the methodologies of teaching physical sciences and to encourage students to study scientific subjects. She is the author of more than 50 research articles, reviews, congress lectures, and other communications and was the corresponding author of an article that received an award from the Italian Research Evaluation Panel.

1 Introduction

Chromatography is probably the most widely used technique in a modern analytical laboratory since it can simultaneously separate and quantify analytes. A wide range of very different samples in the environmental, pharmaceutical, forensic, food, and life science fields can be analyzed by several chromatographic strategies that all share the same principles, based on the differential affinities of individual chemical species for two immiscible phases: the stationary and the mobile phases.

Molecules set in motion by the mobile phase (eluent) move through the stationary phase, suitably immobilized on a medium. The higher the affinity for the stationary phase and the lower the affinity for the mobile phase, the slower the analyte. As in a race, the fastest chemical species cover a prearranged distance in the shortest time, arrive at the "finish line," and produce a detector signal proportional to the amount of analyte. The aggregation state of the mobile phase enables us to differentiate liquid, gas, and supercritical chromatographic techniques.

Chromatography was invented in 1901 by Tswett, a Russian botanist, who separated plant pigments via column liquid chromatography with calcium carbonate as the stationary phase. He first used the *chromatographic method* term in print in 1906 [1]. High performance liquid chromatography (HPLC) is a modern translation of Tsvet's column liquid method. HPLC is no longer very technologically demanding and expensive. Tremendous improvements achieved in recent decades allow experienced and capable practitioners to exploit its formidable potential. Most separations are performed on apolar stationary phases with polar hydro-organic eluents: this mode is called reversed phase high performance liquid chromatography (RP-HPLC) since the polarity of the chromatographic bed is "reversed" unlike the hydrophilic beds in the early techniques.

One of the biggest challenges facing chromatography users is the separation of ionic chemical species for which the use of RP-HPLC is somewhat restricted because they are barely retained on the apolar stationary phase. pH adjustment of the mobile phase to suppress analyte ionization is only suitable for simple mixtures of analytes with similar pK_a values that lie in the pH range of stability of the chromatographic bed (pH of 2 to 8 for silica-based stationary phases). Ion exchange chromatography is an alternative separation method for ionic solutes; it is performed on stationary phases bearing ionized groups oppositely charged to the ionic analyte. It follows that its selectivity is limited because the hydrophobic moiety of the charged species do not strongly contribute to the analyte retention driving force.

The drawbacks of ionic suppression and ion exchange chromatography led to the development of ion-pair chromatography (IPC)—an intriguing mode of HPLC that allows the separation of complex mixtures of polar, ionic, and ionogenic species. IPC is now an established and valuable separation strategy.

The IPC technique is not a passing fad. It represents an advance that will have a lasting impact based on its prominent place in the HPLC armory for tackling difficult analyses of ionogenic solutes. Its scope is substantial, because organic and inorganic ions, ionogenic, neutral, and zwitterionic compounds may be analyzed via typical or modified IPC approaches. The main purpose of this book is to expand the present status of IPC. It is important to underline that this book represents a trade-off between breadth and depth. Many recent reports on a variety of aspects of IPC research are selected for inclusion in this book to provide an up-to date overview of this separative method. The progress from theoretical issue to application will be illustrated in detail, to critically gather most of the representative aspects of this topic into a special volume that may be used as a textbook without compromising its research-oriented character.

REFERENCE

1. Berezkin, V.G., Ed., *Chromatographic Adsorption Analysis: Selected Works of M.S. Twett*. Ellis Horwood: New York, 1990.

2 Electrolyte Solutions and Historical Concept of Ion-Pairing

2.1 PHENOMENOLOGICAL TREATMENT

What is an ion-pair? In order to shed light on the theory that governs ion-pairing, the process first will be treated phenomenologically.

In a naïve picture, an ion-pair may be described as a couple of oppositely charged ions temporarily held together by electrostatic attraction, without establishment of a durable chemical bond. An ion-pair is a diverse chemical species, but the distinction between ion-pairs and free ions that interact via long-range non-specific ion–ion forces is questionable. As a rule, species are regarded as ion-pairs if the distance (r) between two oppositely charged ions bathed in solution is lower than a cut-off length (R). Obviously the ion-pair partners cannot approach each other more closely than the distance of closest approach (a) because of the strong repulsive interactions of their electron shells. The ions are deemed to be paired if their distance apart is between a and R. Moreover the lifetime of the ion-pair should be at least longer than the time required to thermally diffuse over such distance, since a simple encounter of a cation and an anion does not involve ion-pair formation. Lifetimes as low as ~1 ns have been reported [1].

When each ion maintains its own primary solvation shell, the new chemical species is a solvent separated ion-pair (SSIP). If a single solvent layer is shared by ion partners, the species is a solvent shared ion-pair (SIP). If the cation and the anion are in contact and no solvent molecules are present between them, the form is contact ion-pair (CIP) or intimate ion-pair. Figure 2.1 illustrates multistep ion-pair formation equilibrium. What sets ion-pairing apart from complex formation is the absence of directional covalent coordinative bonds resulting from a Lewis base–acid interaction and a special geometrical arrangement.

The effects of ion-pairing have been noted in countless experimental situations involving conductometry, potentiometry, spectroscopy, solvent extraction, separative techniques, activity measurement, and kinetic behavior among others. As far as chromatography is concerned, the electrical neutrality and the increased lipophilicity of ion-pairs, compared to unpaired ions, are features of utmost importance involved in retention adjustment.

The physics of electrostatics in aqueous solution has attracted scientists' notice for centuries. At present, the solvation of ions, volumes and radii of ions in solution, and ionic interactions are still hotly debated research fields. Recently, thermodynamic, transport and structural data were mutually employed for gaining fruitful physicochemical insights into electrolyte solutions [2]. This chapter recapitulates the essential

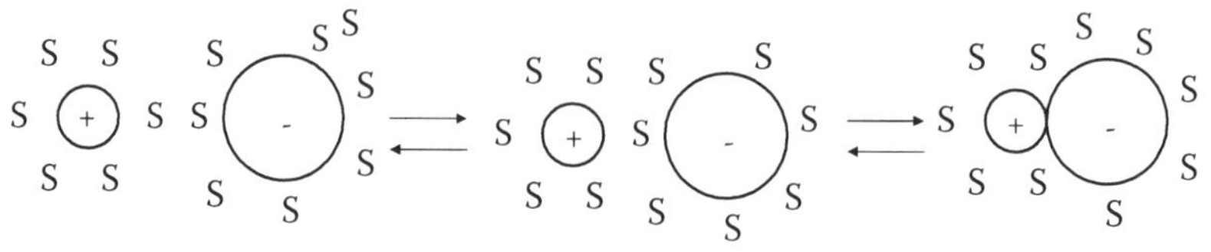

FIGURE 2.1   Multistep ion-pair formation equilibrium.

concepts and considerations concerning ion-pair formation in solution. No attempt is made to be comprehensive when discussing the physical and chemical foundations of the electrolyte solution theory. The interactions of ionic solutes in aqueous electrolytic solutions are discussed in turn: from the ion–solvent interactions that take place even at infinite dilution through ion–ion interactions in diluted and concentrated solutions. While inorganic ion-pair formation has been the subject of many classic treatises on electrolyte solutions and monographs [3–5], a dearth of material covers pairing of organic ions [6,7]. Hence we will carefully ponder the peculiarities of organic ion-pairs since IPC is usually concerned with ions characterized by hydrophobic moieties. For the sake of brevity, the discussion will focus on ion-pair architectures of chromatographic interest, that is, on symmetrical ion-pairs of univalent electrolytes. Similarly, the descriptions of triple and multiple ions, whose existence is claimed only in very low permittivity solvents, are beyond the scope of this chapter.

2.2   ION–SOLVENT INTERACTIONS

If two point charges q_1 and q_2 are separated by a distance R in a vacuum, they are prone to attractive electrostatic forces, and the electrostatic energy of their interaction can be obtained by Coulomb's law:

$$E = \frac{q_1 q_2}{4\pi\varepsilon_0 R} \tag{2.1}$$

where ε_0 is the vacuum permittivity ($8.854 \cdot 10^{-12}$ J^{-1} C^2 m^{-1}). When a crystalline salt dissolves in a solvent, work must be made to separate the ions to distant places from their sites in the lattice, and the lattice energy is lost. The ions create their own electrical field which acts upon and orients the dipolar solvent molecules, which in turn shield the electrical charges, thereby reducing their net magnitude. It follows that the energy of interaction is reduced by a factor called the dielectric constant of the medium or relative permittivity ε_r, that is characteristic of each medium.

If the distance R is much higher than the dimension of the molecules of the medium, the latter can be regarded as a uniform material, and the energy of the electrostatic interaction is obtained by the following expression:

$$E = \frac{q_1 q_2}{4\pi\varepsilon_0 \varepsilon_r R} \tag{2.2}$$

The magnitude of ε_r depends on the nature of the liquid: the higher the liquid polarity, the higher the liquid ε_r, and the higher the ability of the liquid to solubilize ionic solids because electrostatic attractions between the cations and anions of the salt are reduced. The dielectric constant of water is very high due to the strong degree of spatial correlation between dipoles of neighboring molecules. This explains why water is by far the most important polar solvent and, in respect to the range of solute that will dissolve in it, is one of the strongest solvating liquids. Since most IPC eluents are hydro-organic electrolyte mixtures, we will focus on the structures of aqueous solutions.

Water is a mixture of short-lived ice-like water clusters in a sea of weakly hydrogen bonded (or non-hydrogen bonded) molecules. The best known model of water approximates the hydrogen bond network as a random, continuous population of bond distances and orientations that naturally exhibits a roughly bimodal distribution with respect to hydrogen bond angle: a two-state mixture of linear and bent hydrogen bonds with the latter much weaker than the former [8, 9]. This raw picture of water structure is supported by the band shape of the O–H stretch [8] since the low frequency component has been attributed to the ice-like population of water molecules, while the high frequency component has been assigned to non-hydrogen bonded molecules. The primary effect of solutes on water is the change in the relative distribution of water molecules in the linear versus bent hydrogen bond conformation [10].

We will follow a structural approach to describe ionic hydration, since, at variance with a pure thermodynamic description of the phenomenon which detects the overall level of association, it gives information on the way water molecules are arranged around the ion.

If the charge to radius of the ion is sufficiently high, ion–water–dipole solvent interactions may be strong enough to break the genuine hydrogen bonded structure of water and a loss in water structure energy may occur. However the high charge density of these ions enforces linear hydrogen bonding in the solvating molecules so that these ions are known as structure making or kosmotropic ions because their net effect is to strongly order solvent molecules through their charge. When the energy of ion–dipole attractive interactions overcomes the losses of lattice and water structure energies, a net hydrophilic hydration energy results [10].

Another mode of structure making is related to the hydrophobic hydration concept [11]. It is usually shown by large ions with centrally localized charges, such as long chain tetraalkylammonium ions—typical ion-pair reagents (IPRs) commonly used in IPC. The alkyl chains force the nearest water molecules to bind more tightly into the water structure beyond them [12] and molecules with four hydrogen bonds (ice-like cages) are more abundant than lower hydrogen bonded ones. Moreover, water structure disruption by thermal agitation is prevented and the lives of the ice-like clusters are increased. The hydrogen bond network is fortified and the water structure is actually strengthened.

The enhancement of water structure for larger tetraalkylammonium cations in solution was confirmed by neutron diffraction studies [13]. The density of this "expanded water" is obviously decreased, compared to that of bulk water [14–16]. The structure making properties of tetraalkylammonium salts were demonstrated to not begin appreciably until the chain reaches the size of a propyl group and the water structure enforcement around an aryl group is similar to that around a butyl group

[17,18]. Lipophilic ions tends to gather together at the surface of water, and hence are surface-active species that lower the surface tension of the solvent, at variance with small and strongly hydrated kosmotropic ions.

Conversely, the behavior of large polarizable inorganic ions is often chaotropic; they are prone only to break the water structure since the ion–dipole interactions are not strong enough to order solvent molecules around the ion [10]. It would be highly desirable to arrange a gamut of chaotropicity. At least three parameters can be used as indicators. First, Gurney [19] put forward the idea that the viscosity B coefficient of the Dole-Jones equation (that describes change in viscosity upon salt addition):

$$\frac{\eta}{\eta_0} = 1 + Ac^{1/2} + Bc \tag{2.3}$$

is able to quantify the effects of the ions on the structure of the solvent, η being the viscosity of a c molar solution and η_0 being the viscosity of the solvent. A is an electrostatic term and B is a measure of the strength of the ion–water interaction. The contributions of the cation and anion to the electrolyte B coefficient are assumed to be additive. A resulting large positive value of B designates the ion as structure making, while a low or even negative value of B is typical of structure breaker ions. As regards aqueous solutions of classical IPRs such as tetraalkylammonium ions, the large increase of viscosity and decrease of conductance were attributed to the ice-like cage formation that results in a larger moving entity. Tetrapropylammonium and tetrabutylammonium ions are excellent structure makers, while tetramethylammonium is a structure breaker and the two effects annul for tetraethylammonium [20].

The second indicator of kosmotropicity is the standard molar entropy of hydration. For all ions it is highly negative; the higher its absolute value, the more water is ordered upon ionic hydration, and the higher the electrolyte kosmotropicity [2,21].

The third parameter that was found valuable for constructing a kosmotropicity scale is the standard molar Gibbs energy of transfer from light to heavy water; since the hydrogen bonds in D_2O are stronger than those in H_2O [10], chaotropic electrolytes are characterized by large positive standard molar Gibbs energy of transfer [2]. When ions are arranged in order of increasing chaotropicity (and decreasing salting-out ability) we have the following series [10,22–24]:

$$PO_4^{-3} < SO_4^{-2} < H_2PO_4^- < HCOO^- < CH_3SO_3^-$$
$$< Cl^- < Br^- < NO_3^- < CF_3COO^- < BF_4^- < ClO_4^- < PF_6^- \tag{2.4}$$

$$Ca^{2+} < Mg^{2+} < Li^+ < Na^+ < K^+ < NH_4^+ \tag{2.5}$$

The scale is currently known as the Hofmeister (or lyotropic) series to honor Franz Hofmeister, who described the effects of salts on a variety of physiological samples and ranked salts in a sequence that later proved universal. The series of papers he wrote more than a century ago were recently translated into English [22]. Hofmeister simply introduced the concept of ranking salts based on common cations and anions. In this context, it is pertinent to underline that Hofmeister categorically stated that the effect of a salt depends on both its anion and cation. Hofmeister's results were subsequently extrapolated to produce the ion-specific series not connected to salt-specific effects. This may explain subtle discrepancies of the series found by different researchers who used different counterions for a given ion.

Based on the position of an ion in the Hofmeister series, it is possible to foretell the relative effectiveness of anions or cations in an enormous number of systems. The rank of an ion was related to its kosmotropicity, surface tension increments, and salting in and salting out of salt solutions (see below) [25]. A quantitative physical chemistry description of this phenomenon is not far off. Molecular dynamics simulations that considered ionic polarizability were found to be valuable tools for elucidating salt effects [26,27].

Another way to look at ion–solvent interaction is to follow a volumetric approach and assign a hydration number to the ion [28–32]. Mathematically speaking, the coulombic force originating from a point charge becomes zero only at infinity; in effect, however, the force fades to a negligible value after a very short distance beyond which the solvent molecules may be regarded as unaware of the ion's presence. The solvation number of an ion can be defined as the effective number of solvent molecules that are permanently aligned in its force field (via ion–dipole interactions) and surrender their own translational freedom, thereby following the movements of the ion from site to site. These solvent molecules are said to be "electrostricted." The solvation number clearly increases with increasing ionic valence and decreases with increasing ionic size. The main drawback of this methodology is the fact that this descriptor of ionic solvation can only be defined operationally, since its value depends on the strategy used for its measurement and it does not have a universal physical meaning.

2.3 ION–ION ASPECIFIC INTERACTIONS

While at infinite dilutions only ion–solvent interactions occur and electrolyte solutions behave ideally, also at very low concentrations they deviate from ideality because of the electrostatic interaction energy of ions. Attractive forces between oppositely charged analytes lower the active concentration of each ionic species, because the attraction changes the way a given ion reacts chemically. Chemical laws are obeyed only if concentration is replaced by another physical quantity, the activity that is proportional to the concentration by a factor known as activity coefficient γ.

The seasoned Debye-Hückel (D-H) theory, put forth in 1923 [33,34] takes into account the contribution of the ionic electrostatic interactions to the free energy of a solution and provides a quantitative expression for the activity coefficients. The basic concept of the D-H theory is that the long-range Coulomb interaction between two individual ions bathed in a salt solution is mediated by mobile ions from the solution. The effective charges of a certain ion are decreased as the result of charge screening by the mobile counterions; it follows that, at sufficient distance, the interaction between two ions decays exponentially. We briefly outline the main considerations and assumptions of the D-H model:

1. Ions are considered rigid, unpolarizable point charges in a dielectric continuum.
2. The presence of solutes does not influence the dielectric constant of a solvent.

3. Forces other than coulombic ones are not taken into account.
4. Each ion is surrounded by oppositely charged ions and solvent molecules that form the co-sphere around the central ion.
5. The counterions cannot get closer than a certain distance to the central ion; this is called the distance of closest approach (a).
6. The energy of ionic interactions is considered to be smaller than the thermal energy because ionic species do not form a new chemical species.
7. The linear approximation of the Boltzmann distribution that allows one to calculate the time averaged density of opposite charge surrounding the central ion, is assumed to hold, and this is actually true only for very low ionic concentrations.
8. At a given concentration, the approximation gives better results the higher temperature and the dielectric constant, and the lower the ionic charge.

The theory uses the Poisson equation to describe the global potential produced by both the central ion and its ionic cloud as a function of radial distance. The potential due to the ionic atmosphere is used to calculate the electrostatic energy for the interaction of the central ion and its atmosphere. If we consider a hypothetical initial state in which ions are not electrostatically interacting (the magnitude of their charge is assumed to be zero) and a final state in which ionic interactions are operating, the free energy change for 1 mol of a certain ion can be easily determined by multiplying the electrostatic energy for the interaction of the central ion and its atmosphere times Avogadro's number N and dividing by two to avoid considering each ion twice (once as a central ion, once as a component of the atmosphere).

This free energy change ($\Delta\mu_e$) represents the electrostatic contribution to the chemical potential of the ion, that is, the electrical work necessary to "charge" the ideal solution, and it is responsible for deviations of the solution from ideal behavior. The activity coefficients of single ions are not measurable experimentally [35]; for an electrolyte E_pH_q, the medium activity coefficient is

$$\gamma_{\pm} = \left(\gamma_+^p \gamma_-^q\right)^{\frac{1}{p+q}} \tag{2.6}$$

and the theory sanctions the following expression:

$$\log\gamma_{\pm} = \frac{-A\,|z_+z_-|\,\sqrt{I}}{1 + Ba\sqrt{I}} \tag{2.7}$$

where A and B are two constants that depend on the solution temperature and solvent dielectric constant, z^+ and z^- are the charge numbers of the cation and the anion of the E_pH_q salt and I is the ionic strength of the solution. The value of a that gives the best fit of results is the mean diameter of the hydrated ions.

The D-H model is considered reliable for concentrations up to 10 mM even if its theoretical basis breaks down below I = 0.001 M. For ionic strengths up to 0.2, the following expression takes into account the self-salting out of the electrolyte due to its ion salvation. If the salt concentration increases, the activity coefficient first decreases, passes a minimum, and then increases [37]; this behavior is described by the following expression in which C is an adjustable parameter:

$$\log \gamma_{\pm} = \frac{-A \, |z_+ z_-| \sqrt{I}}{1 + Ba\sqrt{I}} + CI \tag{2.8}$$

The activity coefficients could also be modelled very correctly by an alternative expression put forth by Pitzer and co-workers, over wide concentrations and temperatures [37].

As far as mixed strong electrolyte solutions are concerned, Harned's rule [38] holds. At constant ionic strength, the activity coefficient of one electrolyte (A) in the mixture is a function of the fractional ionic strength ($y_B I$) of the other electrolyte (B):

$$\ln \gamma_{\pm A} = \ln \gamma_{\pm A}^0 - \alpha_A y_B I \tag{2.9}$$

where α is an interaction parameter that depends on the temperature and the ionic strength and $\gamma_{\pm A}^0$ is the mean ionic activity coefficient of A in the absence of B and at the same ionic strength of the mixture. In some cases a quadratic term is needed to address deviations.

2.4   INTERACTIONS OF IONS WITH NON-ELECTROLYTES

For non-electrolyte solutes in a solution containing electrolytes, salting effects may be significant. The addition of a salt to an aqueous solution containing a neutral chemical species tends to decrease its solubility; this corresponds to an increase of the activity coefficient of the non-electrolyte and this effect is known as salting out. The reverse behavior, known as salting in, is less common. In a naïve picture, salting out results from a decrease of the number of "free" water molecules because, in an electrolyte solution, some free water molecules suffer solvational interactions with ions; their activity is decreased and this results in increased concentration of the neutral solute over its stoichiometric value. Salting in is rarer, but it is impossible to assign exclusive salting-out or salting-in properties to a given electrolyte because salting effects depend also on the nature of the non-electrolyte.

Empirically, the Setchenov equation [37,39] has been found to express the variation of the neutral solute activity coefficient (γ_N) with the electrolyte concentration (c_E), at least for low electrolyte concentrations (a few tens molar):

$$\log \gamma_N = k_N c_E \tag{2.10}$$

or

$$k_N = \lim_{c_E \to 0} \frac{d \log \gamma_N}{dc_E} \tag{2.11}$$

k_N is the salting constant or Setchenov coefficient and may be positive (negative) if salting out (salting in) occurs; it is characteristic of the electrolyte and the neutral solute. Many theories tried to relate this coefficient to parameters of known physical meaning. The solvent was originally treated as a continuum with a uniform dielectric constant. Among electrostatic theories proposed to calculate the salting constant, in the treatment by Debye and McAulay [40], salting effects are related to the change of the relative permittivity of the medium when electrolytes and non-electrolytes are added to a pure solvent. They calculated the difference in electrical work required to charge the ions in water and in the non-electrolyte solution. The relative permittivity of a solution containing both non-electrolytes and electrolytes can be calculated as:

$$\varepsilon_{r,mix} = \varepsilon_r(1 - an_N - bn_S) \tag{2.12}$$

where $\varepsilon_{r,\,mix}$ is the relative permittivity of the solution, a and b are constants, n_N and n_S are the numbers of non-electrolyte and electrolyte molecules, respectively. If a is positive (negative) the macroscopic relative permittivity of the solution is decreased (increased) by the addition of the electrolyte and since γ_N is increased (decreased) salting out (salting in) is predicted. Salting in is predicted for neutral solutes more polar than water characterized by negative values of the a constant. From the difference in the electrical work required for charging the ions in water and in the non-electrolyte solution the following expression can be obtained [35]:

$$k_N = \frac{Nae^2}{4606\varepsilon_r^2 RT} \sum \frac{v_i z_i^2}{r_i} \tag{2.13}$$

where $v_i = c_i/c_E$ and c_i, r_i, z_i are, respectively, the concentration, the radius, and the charge number of the species of type i, N is Avogadro's number, and e is the unit charge. However, predictions by this theory were often neither quantitatively nor qualitatively in agreement with experimental evidence because of the shortcomings related to the use of the Born equation in a very polar medium such as water [39].

Electrostatic theories improved via a treatment in which the distribution of the neutral solute is evaluated around the ions of the electrolyte and compared with that around solvent molecules. The electrostatic polarization energy suffered by the neutral solute, because of the ion electrostatic field, generates a Boltzmann relation that sanctions the salting effects [41].

When the Boltzmann factor was not linearized and dielectric saturation effects, important near the ion, were considered, better results could be obtained also because the specificity of ion–solvent interaction was accounted for via the introduction of a parameter that measured an effective radius of the primary solvation shell where dielectric saturation occurs [42]. Subsequently, the discontinuous nature of the solvent near the ion was successfully taken into account [43].

Long and McDevit's thermodynamic "electrostriction" theory [44] tries to calculate it from the excess work necessary to put an element of volume of the neutral solute into the electrolyte solution instead of into water; the treatment sanctions the following equation

$$k_N = \frac{\bar{V}_N^0 \left(V_S - \bar{V}_S^0 \right)}{2.303 \beta_w RT}$$

(2.14)

where $\bar{V}_N^0$ is the standard partial molar volume of the non-electrolyte at infinite dilution, V_S is the molar volume of the hypothetical super-cooled liquid salt, that is, the intrinsic volume of the electrolyte, $\bar{V}_S^0$ is the standard partial molar volume of the electrolyte at infinite dilution, and β_w is the water isothermal compressibility. Since standard partial molar volume of a species in a mixture is the volume variation when one mole of that species is added to a large volume of the mixture, this quantity is negative for many electrolytes due to compression of the solvent in the presence of ions. This phenomenon is called electrostriction of the solvent and results in a higher internal pressure that reduces the available space for the non-electrolyte, pushing it out of the liquid phase. Salting-in behavior would demand a negative numerator in Equation 2.14. This method often gives estimates of k_N much larger than the observed values; [37].

At variance with electrostatic theories, treatments based on scaled particle theory [45] do not consider electrostatic interactions and structural modification since the ionic solution is regarded as another kind of solvent; the Setchenov coefficient is related to the free energy change for the formation of cavity large enough to hold the non-electrolyte and for the non-electrolyte introduction into the cavity. The model by Masterton and Lee is only suitable for non-polar non-electrolytes and must be customized for polar non-electrolytes, but in this case the complexity of the representation increases considerably.

Statistical mechanical treatments fail to address the dispersion and polarization interactions that are particularly important when hydrophobic reagents such as IPRs are dealt with [46]. The quantitative treatment of these interactions was introduced by Bockris, Bowler-Reed, and Kitchener [47]; their work is important for explaining anomalous salting in when the simple electrostatic theory would predict salting out.

As far as IPC is concerned, relatively large polarizable solutes may be expected to be salted in by large organic ions such as IPRs. Actually [48,49] it was demonstrated that sharing of co-spheres of a hydrophobic non-electrolyte and a hydrophobic ion leads to salting-in effects and mutual salting in is expected for all hydrophobic solutes in water; it was also demonstrated that the increase of ice-like structures of water near large ions (hydrophobic hydration) translates into negative electrostriction due to the decreased density of the ice-like water. This accounts for the success of Long and McDevit (Equation 2.14) in predicting salting in [49].

From a molecular view, the decrease of entropy upon hydrophobic hydration is not mitigated by a large hydration enthalpy and this translates into an increase in the free energy of water. A system will tend to minimize this increase in free energy through association of the hydrophobic moieties. This phenomenon that explains the salting in of a neutral hydrophobic molecule by hydrophobic ions is expected to amplify with the sizes of the hydrophobic moieties [49]. Attractive forces between two hydrophobic ions and repulsive forces between hydrophilic and hydrophobic ions were related, respectively, to structural salting in, eventually leading to ionic association and structural salting out [11].

In Chapter 3 the significant consequences of salting effects on charged and neutral analyte retention in IPC will be treated in detail.

As a concluding remark it is pertinent to observe that proving a theory requires several physical quantities that cannot unambiguously be determined; simple and accurate prediction of salting coefficients should spur further theoretical development because their predictions are only fairly accurate. It is, however, clear that the logarithm of the activity coefficients of solutes in electrolyte solutions contains a term linear in the electrolyte concentration: for a single electrolyte, a mixture of two electrolytes, and a non electrolyte, Equations 2.8, 2.9, and 2.10, respectively, sanction the quantitative relationships.

2.5 CRITICAL REVIEW OF HISTORY OF THEORETICAL TREATMENTS OF ION-PAIRING

2.5.1 FROM THE BEGINNING TO BJERRUM'S MODEL

The history of the ion-pair concept begins in the 1880s [50] with the theory of Arrhenius that described the electrolytic dissociation in solution as a function of electrolyte concentration and nature. This theory was completely eclipsed by the success of the D-H treatment of electrolyte activity in the 1920s [33] that also paved the way to the theoretical development of the conductivity theory [51] without resort to ion association.

Exact D-H experimental predictions break down at low concentrations; hence early conductometric evidence of ion-pairing rapidly sustained the ion-pair concept. Bjerrum, in 1926, endorsing Brönsted's idea of specific ionic interactions [52], suggested an interesting improvement that ascribed all departures from the D-H model predictions to ion association. His theory developed from the consideration that if a counterion in the ionic cloud became sufficiently close to the central ion during random thermal movement, its thermal translational energy would not be sufficient for it to continue its independent movement in solution, since the two ions would be trapped in each other's electric field. Upon colliding, the ions in opposite charge status stay together only for a short time. This way an ion-pair may be formed; the pair concept was formally introduced [53]. Ions were treated as hard spheres pairwise interacting in a dielectric continuum; these set of assumptions are currently known as the "restricted primitive model" (RPM) of electrolyte solution. Bjerrum presumed that the ion-pair formation cut-off distance, R, is equal to

$$q = \frac{z^2 e^2}{2\varepsilon k_B T} \tag{2.15}$$

where z is the charge number of the ions of the symmetrical ion-pair, $\varepsilon = 4\pi\varepsilon_\circ\varepsilon_r$, k_B is the Boltzmann constant, T is the Kelvin temperature, q represents the distance for which the probability of finding a counterion in a spherical shell next to the central oppositely charged ion is minimum. For lower distances, the probability increases because, even if the number of ions in the shell thickness is very low, coulombic attraction is stronger. For higher distances, the probability increases because the volume of the spherical shell and the number of ions in the shell are greater; however the lower electrostatic force leads us to believe that ions are unable stick together, because they are scattered apart by thermal motion and ion-pairs are not formed.

For distances larger than q, the usual D-H treatment applies to ions that are considered free. The cut-off distance is reasonable, because at that separation distance, the thermal energy is half the work necessary to separate the ion-pair. The ions in the couple cannot get closer than a distance of closest approach, a, which is at least the sum of the crystal radii, but it is not larger than the sum of the solvated ion radii. Since the neat charge of the duplex is zero, the ion-pair is not acted upon by coulombic fields, it does not migrate, and it does therefore not contribute to the electrical conductivity of the solution.

However, as an ionic dipole, it may rotate in an external electrical field. On the average, a precise population of these duplexes exists, although the formation and dissociation of ion-pairs are incessant. The degree of association can be easily obtained if the medium dielectric constant and the ion sizes, and the electrolytel concentration (c) are known, since ion-pairing can be treated as an equilibrium reaction between a positive ion (E^+) and a negative ion (H^-), characterized, according to the law of mass action, by a thermodynamic association constant K_{EH}:

$$E^+ + H^- \rightarrow EH \tag{2.16}$$

$$K_{EH} = \frac{a_{EH}}{a_{E^+}a_{H^-}} \tag{2.17}$$

If α is the fraction of free ions and $(1 - \alpha)$ is the fraction of paired ions we have

$$K_{EH} = \frac{(1-\alpha)\gamma_{EH}}{\alpha^2 c \gamma_\pm^2} \tag{2.18}$$

where γ_{EH} and $\gamma_\pm$ are, respectively, the activity coefficient of the neutral ion-pair and of the mean activity coefficient of free ions subject to non-specific electrostatic interactions. The former is often approximately set to unity, while the latter may be obtained by the D-H theory but with the distance of closest approach given by q and the ionic strength given by αc, since it is determined only by the concentration of the dissociated ions. Recently Marcus noted that the generally accepted supposition that $\gamma_{EH} = 1$ is not always tenable [54]. The fraction of paired ions can be obtained by integration of the probability of the i-ion to be in a spherical shell of radius r and thickness dr around the j-ion from a to q. When the solution is so diluted that $\gamma_\pm$ is close to unity and the b parameter is defined as

$$b = 2q/a = \left(\frac{z^2 e^2}{a \varepsilon k_B T} \right) \tag{2.19}$$

the final expression is:

$$K_{EH} = (4\pi N_A/1000)\left(z^2 e^2/\varepsilon k_B T\right)^3 Q(b) \tag{2.20}$$

where $Q(b)$ is an integral calculated by Bjerrum [53]. From this expression it follows that the higher the dielectric constant of the solvent and the lower the ion charge, the less ion-pairing is expected. It can be calculated that for solvents of high relative permittivity and large singly charged symmetrical electrolytes, $q < a$ and consequently they are not expected to be electrostatically ion-paired in water, according to this model.

The major criticism of the Bjerrum theory of ion-pairing, apart from its basis in the RPM description of the electrolytic solution, concerns the uncertainty in the choice of cut-off distance. It is embarrassing to observe that the association distance rises to infinity as the absolute temperature approaches zero. Furthermore, unrealistic predictions of the distance of closest approach follow from the theory [55]. It can be speculated that there is no ionic specificity, apart from size effect, and that contributions other than electrostatic that are crucial in a chromatographic system are neglected. We will show that this impairs the ability of the theory to quantitatively predict thermodynamic equilibrium constants for organic ion-pairs involved in IPC.

2.5.2 FURTHER DEVELOPMENTS

The contribution of the coulombic interaction to the formation of stable ion-pairs is by far the most important and for a long time was regarded as the only one. Denison and Ramsey [56] put forth a thermodynamic approach to ion-pairing that considers only ions in contact to be ion-pairs; this is tantamount to note the cut-off distance as $R = a$. A Born cycle was then used to calculate the Gibbs energy of separating ions from contact to infinity and the equilibrium sanctions the following association equilibrium constant

$$K_{EH} = \exp(b) \tag{2.21}$$

Their theory predicts that, for these contact ion-pairs, $\ln K_{EH}$ should be linear with $1/\varepsilon$, but this was at variance with steady accretion of direct evidence [57]. Gilkerson continued to take into account only contact ion-pairs but, for the first time, the role played by solvation, excluded by the continuum model of the solvent, was not ignored [57,58]. The expression for the ion-pairing equilibrium constant can be written in the following terms:

$$K_{EH} = A \exp\left(-E_{sol}/RT\right)\exp(b) \tag{2.22}$$

The term A is related to the solvent density and molecular weight and to the free volumes of the ions and the ion-pair [59]. E_{sol} is the difference between the molar ion-pair solvation energy and the free ion solvation energy. The theory does not predict a simple linearity of $\ln K_{EH}$ with $1/\varepsilon$. Actually solvent effects other than that due to the relative permittivity of the solvent are easily predicted, since the macroscopic ε is only a rudimentary description of the real attenuation of the ionic interactions due to the polarization of the solvent molecules by the intense electric field surrounding the ions. Moreover preferential solvation may occur if the dipole field of one component of the mixture is much larger than that of the other. In this context it is crucial to emphasize that even if solvation energy changes may be negligible for inorganic ions, solvophobic interactions are expected to be significant for large lipophilic ions as classical IPRs.

Fuoss adopted the concept of the electrostatic contact ion-pair [60] and considered the anion as a point charge that may also penetrate the cation-conducting sphere of radius a. The final expression for the ion-pairing equilibrium constant,

$$K_{EH} = \left(\frac{4\pi a^3 N_A}{3000}\right)\exp(b) \tag{2.23}$$

is similar to that developed by Gilkerson. The pre-exponential factor has the meaning of an excluded volume; since it is related only to a, the Fuoss theory predicts that $\ln K_{EH}$ should be linear with $1/\varepsilon$. This prediction, again, was refuted by experimental findings obtained in binary solvent mixtures where ε could be varied continuously by changing the composition. Later, Justice and Justice [61] criticized the fact that the contact arrangement is only a highly improbable configuration compared to those in which the anion–cation distance is lower than a. On the other hand, they emphasized that the penetration of the anion into the cation sphere is implausible due to the short-range repulsive potentials at such distances. Fuoss in a later paper introduced a solvation term [62] in the ion-pairing equilibrium constant, obtaining an expression that is tantamount to that put forth by Gilkerson:

$$K_{EH} = \left(\frac{4\pi a^3 N_A}{3000}\right)\exp(-E_{sol}/RT)\exp(b) \tag{2.24}$$

The advancement of the theoretical description of ion-pairing was marked by the distinction between internal (or contact or tight) and external (or solvent separated, loose) ion-pairs. Eigen and Tamm [63,64] proposed a stepwise formation of the contact ion-pair: while the formation of the solvent separated ion-pair is diffusion-controlled, the elimination of the solvent molecules to form the contact ion-pair was the slowest stage. Ultrasonic absorption data supported the so-called Eigen mechanism represented in Figure 2.1.

Justice and Justice [61] founded their theoretical description of ion-pairing on the Rasaiah and Friedman formulation of the concentration dependence of the activity coefficient. They took into account both a short-range interaction energy E_{sr} and a long-range coulombic term and a many-body interaction was considered. Their final expression is in agreement with that of Bjerrum, since $E_{sr} = \infty$ for $r \le a$ and $E_{sr} = 0$ for $r > a$, and the result [65] is comparable to Bjerrum's.

The model by Justice and Justice was adopted by Barthel. In its low concentration chemical model [66,67] the equilibrium between free ions and ion-pairs is considered and a pair distribution function of a symmetrical electrolyte in solution is determined, avoiding a truncated series expansion of the Boltzmann factor of Coulomb ion–ion interaction in the vicinity of an ion and introducing a mean potential of short-range forces. The cut-off distance was $R = a + nd_s$, where d_s is the diameter of the solvent molecule located between the pairing partners and $n = 1$ or 2 (SIP and SSIP, respectively).

The solvation and correlation are taken into account, thereby departing from the RPM. The dependence of solvation on distance was recently investigated [67]. The expression of the ion-pairing equilibrium constant is analogous to that developed by Gilkerson or by Fuoss. Again, the solvation parameter avoids the linearity of $\ln K_{EH}$ with $1/\varepsilon$.

While in the 1970s, agreement on the coulombic term was achieved, it was not clear how to deal with short-range interactions. It is worth noticing that the latter strongly concur to regulate the chromatographic retention. Byberg et al. [68] allowed for dielectric saturation [69] inside the solvation shells of ions and, for aqueous solutions, the assigned relative permittivity was 5.5. This statement vindicated the likelihood of the pairing of univalent ions in aqueous solution [54]. Dielectric saturation in the high electric field near the ions causes a linear drop (up to 2 N) of the dielectric constant of electrolyte solutions with increasing ionic concentration [70,71]. In the mean spherical approximation (MSA) treatment of the ion association in aqueous solutions, the linearity of the relative permittivity and of the hydrated cation diameters with the electrolyte concentration was taken into account and a good fit of the experimental activity and osmotic coefficient was obtained [72–75]. The MSA model was elaborated on the basis of cluster expansion considerations involving the direct correlation function; the treatment can deal with the many-body interaction term and with a screening parameter and proved expedient for the interpretation of experimental results concerning inorganic electrolyte solutions [67,75–77].

Current opinions concerning ion-pair modelling share wide consensus on the electrostatic description of the pairing interaction (even if the appropriate values of a and R are still debated), but call attention to the repulsion at $r < a$, and to the importance of free solvation energy change and other short-term interactions.

2.5.3 Hydrophobic Ion-Pairing Concept

A critical review of the history of the ion-pair concept indicates that considerations other than electrostatic were scarcely provided by model makers. Clearly, for chromatography, solvophobic interactions usually neglected by ion-pair model makers are crucial. Chromatographers involved in ion-pair strategies realized that the theoretical description of the ion-pair was critical; at the time of the ion-pair chromatography introduction in the late 1970s and subsequently, the electrostatic model of ion association dominated scientific debate.

Diamond was the first to focus on the concept of hydrophobic association and demonstrated that, at variance with the Bjerrum theory, ion-pairing of univalent organic electrolytes in water is possible [12]. He capitalized on the hydrophobic hydration concept [11,12] typical of large organic ions (*vide supra*) that increase the water structure via the formation of ice-like cages, thereby decreasing the system entropy and increasing its free energy. The system will therefore tend to minimize this increase in free energy through association of the hydrophobic moieties. This phenomenon termed water structure-enforced ion-pairing (or hydrophobic ion-pairing) is characterized by the opposite needs of electrostatic ion-pairing since it amplifies with the sizes of the ions and can occur only in water or highly structured solvents [12,49].

The cation and anion are forced together by their mutual affinity and by a solvophobic driving force that enforces the coulombic electrostatic attraction and contributes to the formation of the ion-pair. The capability of water to force poorly hydrated ions together so as to decrease their modification of the water structure decreases with increasing percentages of the organic modifier. This again runs counter to the electrostatic description of the pairing process that predicts an improved association with decreasing relative permittivity of the solvent.

Thomlinson [78] was the first chromatographer to point out that the classical electrostatic ion-pair concept did not hold for IPRs that were usually bulky hydrophobic ions; he also emphasized that in the interfacial region between the mobile and the stationary phases, the dielectric constant of the medium is far lower than that of the aqueous phase. Chaotropes that break the water structure around them and lipophilic ions that produce cages around their alkyl chains, thereby disturbing the ordinary water structure, are both amenable to hydrophobic ion-pairing since they are both scarcely hydrated. The practical proof of such ion-pairing mode can be found in References 80 and 81; many examples of such pairing modes are reported in the literature [79–86].

The following experimental evidence confirms that a simple electrostatic modelling of the hydrophobic ion-pairing process is not predictive in its own right. First, it was experimentally observed that, for an organic anion, ion-pairing equilibrium constants in aqueous solutions are higher (lower) if the counterion is organic (inorganic), but the pairing constants of an ion-pair characterized by an organic anion and an inorganic cation are not negligible [79,82]. Second, the ion-pairing equilibrium constant increases with increasing organic ion size [18,83–86]. These observations run counter to the genuine electrostatic dependence of the pairing constant on ion size [79]. Conversely, the trend of the ion-pairing equilibrium constant values parallels the increasing enforcement of the hydrogen bonded structure of water [18] or the increasing mutual affinity of the organic chains of the ions. This stacking free energy, E_{st}, that characterizes the interactions of hydrophobic moieties of the cation and the anion can be considered a kind of adsorption energy that is usually linearly dependent on the concentration of the organic modifier [87].

Results by Cecchi and co-workers quantitatively indicated that ion-pairing was favored in water-rich eluents, again at variance with the genuine electrostatic theory predictions [87]. These results are confirmed by the switch from salting-in to salting-out effects of tetraalkylammonium salt on benzene when water was replaced by methanol [88]. It is now generally accepted [7] that organic solvents, at variance with the electrostatic description of the association process, may also run counter to ion-pairing if the solvation is strong: in this case a trade-off exists between ion solvation and ion association since the latter involves a sequential release of solvent molecules from the ionic solvation shell in the region between the partner ions. The ions are still solvated outside this space, and the larger and bulkier the ion, the stronger these solvation effects are predicted to be. The theoretical modelling of the ion-pairing process has a strong effect on the theoretical modelling of the IPC retention mechanism. This issue will be discussed in further detail in Chapter 3.

2.6 THERMODYNAMIC PROPERTIES OF ION-PAIRS

From the thermodynamic equilibrium constant, the estimate of the standard molar free energy of ion-pair formation is sanctioned by the well known relationship:

$$\Delta G_{EH}^0 = -RT \ln \left(K_{EH}/M^{-1} \right) \tag{2.25}$$

The heat of dissociation of an ion-pair may be experimentally estimated calorimetrically from the heat of dilution. This method presents some difficulties because at very high dilution it needs extrapolation procedures, while at higher concentrations, it is hard to distinguish electrostatic and pairing effects. A more straightforward method is study of the van't Hoff plot that represents the expression below that follows from Equation 2.25:

$$\ln K_{EH} = -\frac{\Delta H_{EH}^0}{RT} + \frac{\Delta S_{EH}^0}{R} \tag{2.26}$$

When $\ln (K_{EH}/M^{-1})$ is plotted against T^{-1}, ΔH_{EH}^0 and ΔS_{EH}^0 can be easily obtained, respectively, from the slope and the intercept of the linear relationship that describes the influence of temperature on the ion-pairing equilibrium constant.

ΔH_{EH}^0 is positive if the enthalpy of desolvation of free ions (endothermic process) involved in the duplex formation is not compensated by the enthalpy of the ionic electrostatic attraction (exothermic process). Positive values of ΔS_{EH} are common, because upon ion-pairing, a number of solvent molecules from the solvation shell located in the region between the ions are released to the bulk solvent. A quantitative estimate of this number is possible [89]. Usually this entropy increase outweighs the small entropy decrease because two ions form one single particle, the ion-pair. The extent to which the one factor overwhelms the other depends on the nature of the ions, the strength of their solvation, and the extent to which desolvation occurs during the multi-step formation of the ion-pair.

Ion pairing is also predicted to produce a positive volume change since the sum of the standard partial molar volumes of the free ions is lower than the standard partial molar volumes of the ion-pair. The difference between these two quantities can be obtained by the pressure derivative of K_{EH} (corrected for isothermal compressibility if the molar scale is used) [7]. Solvation effects give reasons for this outcome: the strong electric field of the free ions causes electrostriction of the solvent molecules whose molar value is lower than the bulk one. When the field is decreased, because of the dipolar species formation, the release of some electrostricted molecules is associated with a volume expansion that offers an additional method (*vide supra*) to estimate the number of solvent molecules released upon ion-pairing [31].

2.7   TECHNIQUES FOR STUDYING ION-PAIRING

Many methods were used to obtain qualitative and quantitative data concerning ion-pairing [90,91]. We will discuss the most significant ones.

2.7.1   CONDUCTOMETRY

Conductometry paved the way for the development of the ion-pair concept [3]. The oldest experimental evidence of ion-pairing was obtained from colligative properties and electrical conductivity measurements. It is generally accepted that electroneutral ion-pairs do not contribute to solution conductivity. Conductometry is now a reliable and well established technique even in low millimolar concentration ranges, but the full description of conductance in the presence of ion-pairing is a non-trivial task. To date the most accepted equation was developed by Fuoss and Hsia [92] and expanded by Fernandez-Prini and Justice [93]:

$$\Lambda = \alpha \left[\Lambda^\circ - S(\alpha c)^{1/2} + E\alpha c \ln(\alpha c) + J_1(R_1)\alpha c - J_2(R_2)(\alpha c)^{3/2} \right] \qquad (2.27)$$

where Λ° is the molar conductivity at infinite dilution, and S, E, J_1, and J_2 are known expressions concerning relaxation and electrophoretic effects; R_1 and R_2 are ion distance parameters (usually $R_1 = R_2 = q$). The fit of experimental data gives the best estimate of α, from which the association constant may be obtained via Equation 2.18 if appropriate expressions for activity coefficients are used [94]. The main drawback of conductometric quantification of ion-pairing is that the reliability of the model decreases as the ion-pair association constant decreases.

2.7.2   POTENTIOMETRY

Potentiometric methods have eliminated the problems that beset earlier studies, due to the high electrolyte concentrations required for ideal electrode behavior. Following the so-called constant ionic medium principle [91], a large excess of an indifferent (or inert or swamping) electrolyte is added, so that the activity coefficients of the species can be considered constant when their concentration (very low compared to that of the indifferent electrolyte) are changed over a wide range.

Electrodes sensitive to one of the ion-pair partners in the so-called constant ionic strength cell [95] proved to be valuable to measure the free ion concentration and to determine the stoichiometric equilibrium constant. The latter has a clear thermodynamic meaning if the ionic strength of the medium is indicated, since in this approach, the reference standard state is not the usual infinite dilution of all species dissolved in the solvent ($\gamma \to 1$, as $c \to 0$), but is the infinite dilution of the reacting species in the constant ionic medium ($\gamma \to 1$, as $c \to 0$ at I = constant) [7]. Even if the constant ionic strength attenuates the variation of liquid junction potentials, the lower the association constant, the lower the consistency of the obtained constant.

2.7.3   Spectroscopic Methods

Ion pairing study via regular spectroscopic techniques (UV-vis, IR, Raman, NMR) does not involve major theoretical development, since ion-pairing is detected by noticing modifications of the spectra of free ions. When the absorbance of an electrolyte solution is measured in the UV-vis range, and apparent deviation from the Lambert-Beer law is attributed to ion-pairing, we can write for unit cell length:

$$A = (\varepsilon_{E+} + \varepsilon_{H-})\alpha c + \varepsilon_{EH}(1-\alpha)c \qquad (2.28)$$

where ε_{E+}, ε_{H-}, and ε_{EH} are the absorption coefficients of each species and c is the stoichiometric concentration of the salt. When suitable wavelengths are selected, α and hence the ion-pair equilibrium constant can be easily obtained [3]. Usually only contact ion-pairs may be detected; if a system is characterized by a multi-step ion-pairing reaction, results from typical spectroscopic techniques may be ambiguous [96,97].

2.7.4   Activity Measurement

The fraction of the free ions α, for a 1:1 electrolyte, was shown by solution thermodynamics to be [37, 98]:

$$\gamma_{\pm}^{obs} = \alpha\gamma_{\pm} \qquad (2.29)$$

where $\gamma_{\pm}^{obs}$ is the observed value of the mean activity coefficient that is obviously lowered by ion-pairing. If $\gamma_{\pm}$ is estimated by a theoretical approach and $\gamma_{\pm}^{obs}$ is measured by any common method (such as colligative properties, potentiometry, isopiestic equilibration, among others), α and hence the ion-pair equilibrium constant can be easily obtained, even if the accuracy of the estimates of very low constants is questionable.

2.7.5   Relaxation Methods

These strategies capitalize on the perturbation of an equilibrium system and the measure of the return of the system to equilibrium; they are valuable because they apply almost universally to solutions and can shed light on the multi-step association process. Unfortunately they are technologically demanding and expensive even when commercially available. Additionally, only very experienced and capable practitioners can exploit the potential of these strategies since ion-pairing is detected by the observation and appraisal of new features in spectra.

2.7.5.1   Dielectric Relaxation Spectroscopy (DRS)

The electrolyte solution is stimulated by an electromagnetic field applied over the microwave region, and the dielectric response of the sample is then measured. The complex dielectric response is frequency sensitive and dependent on the square of the dipole moment of each species in solution. Both solvent molecules and dipolar ion-pairs contribute to the signal. Additional contributions arise from the polarizability of the species and from the ions. These components are more important at higher and lower frequencies, respectively. What sets DRS apart from other methods is its great sensitivity to very weakly associated ion-pairs [32] and its ability to distinguish between contact and solvent separated ion-pairs. For the latter, the dipole is higher and the DRS sensitivity is better.

On the other hand DRS is complex and fraught with potential problems, for example, under the effect of the applied field, there is ionic migration and the ionic conductivity term must be deconvolved from the dipolar response at lower frequencies. Moreover DRS is technologically demanding and expensive and only well trained personnel can exploit its potential [99,100].

2.7.5.2 Ultrasonic Relaxation (UR)

A solution is stimulated by sonic waves over the range from audio to hypersonic frequencies (10 kHz to 10GHz) so that adiabatic pressure and temperature perturbations (~1kPa and e ~1mK) result. Usually a spectrum is obtained by plotting sonic energy loss against frequency [101]. The relaxation time constant depends, among other factors, on the concentrations of the species that can be used to quantify the ion-pair equilibrium constant. UR's value is based on its ability to estimate the rates of the forward and reverse reactions involved in a chemical equilibrium including multi-step ion-pairing.

2.7.6 Separative Methods

Both electrophoretic and chromatographic techniques were used recently to study ion-pairing [103–114]. It was observed that the capillary electrophoretic mobility of large anions decreased with increasing concentration of large lipophilic cations in the background electrolyte (BGE) and the ion association constant was estimated by measuring the consequent shift of the analyte peak [106–109]. For the formation of a neutral ion-pair, the ion-pairing equilibrium constant can be obtained from the plot of the following linear expression [103,104]:

$$\frac{\mu^0}{\mu} = 1 + K_{EH}[H^-] \tag{2.30}$$

where μ^0 and μ are, respectively, the electrophoretic mobilities in the absence and in the presence of the pairing ion H^-. Popa and co-workers thoroughly studied ion-pairing in the BGE in the capillary electrophoresis (CE) separation of diastereomeric peptide pairs [110]. Despite the absence of a hydrophobic surface, a clear separation, according to the analyte hydrophobicity, was observed. Since it improves with increasing IPR concentration and hydrophobicity, the CE method achieves a bi-dimensional separation mode according to the analyte charge and hydrophobicity, thereby demonstrating ion-pairing in solution [111].

It was also more interesting to observe that the value of hydrophobic ion-pairing equilibrium constant runs counter to predictions of a purely electrostatic approach since it increases with increasing size of the pairing ion [107,109,112] for both organic and inorganic analytes [113], thereby clearly confirming that ion association in aqueous solution, at variance with Bjerrum's theory predictions, occurs and it is governed by hydrophobicity. Again, the electrostatic predictions were confuted by the low dependence of the ion-pairing equilibrium constant on the dielectric constant of the medium [103].

The chromatographic estimate of ion-pairing equilibrium constant via IPC [87,114] will be thoroughly detailed in Chapter 3 (Section 3.1.2) and will confirm that separative techniques are particularly valuable for ascertaining the nature of hydrophobic ion-pairing.

REFERENCES

1. Buchner, R. et al. Hydration and ion-pairing in aqueous sodium oxalate solutions. *Chem. Phys. Chem.* 2003, 4, 373–378.
2. Marcus, Y. On the relation between thermodynamic, transport and structural properties of electrolyte solutions. *Russ. J. Electrochem.*, 2008, 44, 16–27.
3. Davies, C.W. *Ion Association.* Butterworth: London, 1962.
4. Marcus, Y. and Kertes, A.S. *Ion Exchange and Solvent Extraction of Metal Complexes*, Wiley: London, 1969, pp. 189–199.
5. Marcus, Y. *Ion Solvation*, Wiley: Chichester, 1985, pp. 218–244.
6. Szwarc, M., Ed. *Ions and Ion-Pairs in Organic Reactions*, Wiley Interscience: New York, 1972, Vol. 1.
7. Marcus, Y. and Hefter, G. Ion pairing. *Chem. Rev.* 2006, 106, 4585–4621.
8. Walrafen, G.E. Effects of equilibrium H-bond distance and angle changes on Raman intensities from water. *J. Chem. Phys.* 2004, 120, 4868–4876.
9. Weinhold, F. Resonance character of hydrogen-bonding interactions in water and other H-bonded species. *Adv. Protein Chem.* 2005, 72, 121–155.
10. Nucci, N.V. and Vanderkooi, J.M. Effects of salts of the Hofmeister series on the hydrogen bond network of water. *J. Mol. Liq.* 2008, 143, 160–170.
11. Desnoyers, J.E., Arel, M., and Leduc, P.A. Conductance and viscosity of n-alkylamine hydrobromides in water at 25°C: influence of hydrophobic hydration. *Can. J. Chem.* 1969, 47, 547–553.
12. Diamond, R.M. Aqueous solution behaviour of large univalent ions: a new type of ion-pairing. *J. Phys. Chem.* 1963, 67, 2513–2517.
13. Turner, J. and Soper, A.K. The effect of apolar solutes on water structure: alcohols and tetraalkylammonium ions. *J. Chem. Phys.* 1994, 101, 6116–6125.
14. Frank, H.S. and Wen, W.Y. Structural aspects of ion–solvent interaction in aqueous solutions: a suggested picture of water structure. *Disc. Faraday Soc.* 1957, 24, 133–140.
15. Conway, B.E., Verrall, R.E., and Desnoyers, J.E. Specificity in ionic hydration and the evaluation of individual ionic properties. *Z. Phys. Chem. (Leipzig)* 1965, 230, 157–178.
16. Wen, W.Y. Structural aspects of aqueous tetraalkylammonium salt solutions. *J. Sol. Chem.* 1973, 2, 253–276.
17. Evans, D.F., Cunninghan, G.P., and Kay R.L. Interaction of tetraethanolammonium ion with water as determined from transport properties. *J. Phys. Chem.* 1966, 70, 2974–2980.
18. Takeda, Y., Ikeo, N., and Sakata, N. Thermodynamic study of solvent extraction of 15-crown-5- and 16-crown-6-s-block metal ion complexes and tetraalkylammonium ions with picrate anions into chloroform. *Talanta* 1991, 38, 1325–1333.
19. Gurney, R.W. *Ionic Processes in Solution*, McGraw Hill: New York, 1953, p. 173.

20. Kay, R.L. et al. Viscosity B coefficient for the tetralkylammonium halides. *J. Phys. Chem.* 1966, 70, 2336–2341.
21. Nightingale, E.R., Jr. Phenomenological theory of ion solvation: effective radii of hydrated ions. *J. Phys. Chem.* 1959, 63, 1381–1387.
22. Kunz, W., Henle, J., and Ninham, B.W. Zur Lehre von der Wirkung der Salze (About the science of the effect of salts): Franz Hofmeister's historical papers. *Curr. Opin. Colloid Interface Sci.* 2004, 9, 19–37.
23. Flieger J. The effect of chaotropic mobile phase additives on the separation of selected alkaloids in reversed-phase high-performance liquid chromatography. *J. Chromatogr. A.* 2006, 1113, 37–44.
24. Kazakevich, I.L. and Snow, N.H. Adsorption behavior of hexafluorophosphate on selected bonded phases. *J. Chromatogr. A.* 2006, 1119, 43–50.
25. Zhou, H. X. Interactions of macromolecules with salt ions: an electrostatic theory for the Hofmeister effect. *Proteins* 2005, 61, 69–78.
26. Jungwirth, P. and Tobias, D. J. Molecular structure of salt solutions: a new view of the interface with implications for heterogeneous atmospheric chemistry. *J. Phys. Chem. B.* 2001, 105, 10468–10472.
27. Horinek, D. et al. Molecular hydrophobic attraction and ion-specific effects studied by molecular dynamics. *Langmuir* 2008, 24, 1271–1283.
28. Hinton, J.F. and Amis, E.S. Solvation numbers of ions. *Chem. Rev.* 1971, 71, 627–674.
29. Ohtaki, H. *Monatsh.* Ionic solvation in aqueous and nonaqueous solutions. *Monatsch. Chem.* 2001, 132, 1237–1268.
30. Marcus, Y. The solvation number of ions obtained from their entropies of salvation. *J. Solution Chem.* 1986, 15, 291–306.
31. Marcus, Y. Electrostriction, ion solvation, and solvent release on ion-pairing. *J. Phys. Chem. B.* 2005, 109, 18541–18549.
32. Padova, J. Solvation approach to ion–solvent interaction *J. Chem. Phys.* 1964, 40, 691–694.
33. Debye, P. and Hückel, E. Theory of electrolytes I: Lowering of freezing point and related phenomena. *Z. Physik 1923, 24, 185–206.*
34. Debye, P.J.W. *The Collected Papers of Peter J.W. Debye.* Interscience: New York, 1954.
35. Marcus, Y. and Kertes, A.S. *Ion Exchange and Solvent Extraction of Metal Complexes.* Wiley: London, 1969, Chap. 1.
36. Guggenheim, E.A. The specific thermodynamic properties of aqueous solutions of strong electrolytes. *Phil. Mag.* 1935, 19, 588–643.
37. Pitzer, K.S. In *Activity Coefficients in Electrolyte Solutions*, 2nd ed., Pitzer, K.S., Ed., CRC Press: Boca Raton, FL, 1991.
38. Harned, H.S. Some thermodynamic properties of uni-univalent halide mixtures in aqueous solution. *J. Am. Chem. Soc.* 1935, 57, 1865–1873.
39. Conway, B.E. Local changes of solubility induced by electrolytes: salting-out and ionic hydration. *Pure Appl. Chem.* 1985, 57, 263–272.
40. Debye, P., McAulay, J. Z. Das Elektrische Feld Der Ionen Und Die Neutralsalzwirking (The electric field of ions and the action of neutral salts). *Physik* 1925, 26, 22–29.
41. Butler, J.A.V. The mutual salting-out of ions *J. Phys. Chem.* 1929, 33, 1015–1023.
42. Conway, B.E., Desnoyers, J.E., and Smith, A.C. On the hydration of simple ions and polyions. *Phil. Trans. Roy. Soc. London A.* 1964, 256, 389–437.
43. Conway, B.E., Novak, D.M. and Laliberté, L. Salting-out and ionic volume behavior of some tetraalkylammonium salts. *J. Solution Chem.* 1974, 3, 683–711.
44. Long, F.A. and McDevit, W.F. Activity coefficients of nonelectrolyte solutes in aqueous salt solutions. *Chem. Rev.* 1952, 51, 119–169.

45. Masterton, W.L. and Lee, T.P. Salting coefficients from scaled particle theory. *J. Phys. Chem.* 1970, 74, 1776–1782.
46. Krishnan, C.V. and Friedman, H.L. Model calculations for Setchenow coefficients. *J. Solution Chem.* 1974, 3, 727–744.
47. Bockris, J.O., Bowler-Reed J., and Kitchener, J.A. The salting-in effect. *Trans. Faraday Soc.* 1951, 47, 184–192.
48. Desnoyers, J.E. and Arel, M. Apparent molal volumes of n-alkylamine hydrobromides in water at 25°C: hydrophobic hydration and volume changes. *Can. J. Chem.* 1967, 45, 359–366.
49. Desnoyers, J.E. and Arel, M. Salting-in by quaternary ammonium salts. *Can. J. Chem.* 1965, 43, 3232–3237.
50. Arrhenius, S. Z. Die Dissociation der im Wasser gelösten Stoffe. (On the dissociation of substances dissolved in water). *Z. Phys. Chem.* 1887, 1, 631–648.
51. Onsager, L. Zur Theorie der Elektrolyte (Electrolyte theory). *Phys. Z.* 1927, 28, 277–298.
52. Brönsted, J.N. Studies on solubility I: solubility of salts in salt solution. *J. Am. Chem. Soc.* 1920, 42, 761–786.
53. Bjerrum, N. Ionic association I: influence of ionic association on activity. *Kgl. Danske Videns. Selskab.* 1926, 7, 1–48.
54. Marcus, Y. On the activity coefficients of charge-symmetrical ion-pairs. *J. Mol. Liq.* 2006, 123, 8–13.
55. Robinson, R.A. and Stokes, R.H. *Electrolyte Solutions*, 2nd ed., Butterworth: London, 1965, p. 422.
56. Denison, J.T. and Ramsey, J.B. Free energy, enthalpy, and entropy of Dissociation of some perchlorates in ethylidene chloride. *J. Am. Chem. Soc.* 1955, 77, 2615–2621.
57. Gilkerson, W.R. Application of free volume theory to ion-pair dissociation constants. *J. Chem. Phys.* 1956, 25, 1199–1202.
58. Gilkerson, W.R. The importance of the effect of the solvent dielectric constant on ion-pair formation in water at high temperature and pressures. *J. Phys. Chem.* 1970, 74, 746–750.
59. Gilkerson, W.R. and Stamm, R.E. Conductance of tetra-n-butylammonium picrate in benzene-o-dichlorobenzene solvent mixtures at 25°C. *J. Am. Chem. Soc.* 1960, 82, 5295–5298
60. Fuoss, R.M. Ionic association III: equilibrium between ion-pairs and free ions. *J. Am. Chem. Soc.* 1958, 80, 5059–5061.
61. Justice, M.C. and Justice, J.C. Ionic interactions in solutions I: association concepts and the McMillan-Mayer theory. *J. Solution Chem.* 1976, 5, 543–561.
62. Sadek, H. and Fuoss, R.M. Electrolyte–solvent interactions VII: conductance of tetrabutylammonium bromide in mixed solvents. *J. Am. Chem. Soc.* 1959, 81, 4507–4512.
63. Eigen, M. and Tamm, K. Schallabsorption in Elektrolytloesungen als Folge chemischer Relaxation I (Acoustic absorption in electrolyte solutions as a consequence of chemical relaxation I.) *Z. Elektrochem.* 1962, 66, 93–121.
64. Eigen, M. and Tamm, K.Z. Schallabsorption in Elektrolytlosungen als Folge chemischer Relaxation II: Messergebnisse und Relaxationsmechanismen fur 2–2-Vertige Elektrolyte. (Acoustic absorption in electrolyte solutions as a consequence of chemical relaxation II: Experimental results and relaxation mechanisms for 2–2-valent electrolyte). *Z. Elektrochem.* 1962, 66, 107–121.
65. Justice, M.C. and Justice, J.C. Ionic interactions in solutions II: theoretical basis of the equilibrium between free and pairwise associated ions. *J. Solution Chem.* 1977, 6, 819–826.
66. Barthel, J. J. Temperature dependence of the properties of electrolyte solutions I: a semi-phenomenological approach to an electrolyte theory including short range forces. *Ber. Bunsen Ges. Phys. Chem.* 1979, 83, 252–257.

67. Krienke, H. and Barthel, J. Association concepts in electrolyte solutions. *J. Mol. Liq.* 1998, 78, 123–138.
68. Byberg, J., Jensen, S.J.K., and Kläning, U.K. Extension of the Bjerrum theory of ion association. *Trans. Faraday Soc.* 1969, 65, 3023–3031.
69. Marcus, Y. and Hefter, G. T. On the pressure and electric field dependencies of the relative permittivity of liquids. *J. Solution Chem.* 1999, 28, 575–592.
70. Hasted, J.B., Ritson, D.M., and Collie, C.H. Dielectric properties of aqueous ionic solutions, parts 1 and 2. *J. Chem. Phys.* 1948, 16, 1–21.
71. Harris, F.E. and Okonski, C.T. Dielectric properties of aqueous ionic solutions at microwave frequencies. *J. Chem. Phys.* 1957, 61, 310–319.
72. Simonin, J.P., Bernard, O., and Blum, L. Real ionic solutions in the mean spherical approximation 3: osmotic and activity coefficients for associating electrolytes in the primitive model. *J. Phys. Chem.B.* 1998, 102, 4411–4417.
73. Blum, L. and Hoeye, J.S. Mean spherical model for asymmetric electrolytes 2: hemodynamic properties and the pair correlation function. *J. Phys. Chem.* 1977, 81, 1311–1316.
74. Wertheim, M.S. Fluids of dimerizing hard spheres, fluid mixtures of hard spheres and dispheres, *J. Chem. Phys.* 1986, 85, 2929–2936.
75. Krienke, H. and Barthel, J. MSA models of ion association in electrolyte solutions. *Z. Phys. Chem.* 1998, 204, 71–83.
76. Barthel, J. et al. The application of the associative approximation in the theory of non-aqueous electrolyte solutions. *Condens. Matter Phys.* 2000, 3, 657–674.
77. Tomšič, M. et al. Conductivity of magnesium sulfate in water from 5 to 35°C and from infinite dilution to saturation. *J. Solution Chem.* 2002, 31, 19–31.
78. Thomlinson, E., Jefferies, T.M., and Riley, C.M. Ion pair high-performance liquid chromatography *J. Chromatogr.* 1978, 159, 315–358.
79. Fini, A.et al. Formation of ion-pairs in aqueous solutions of diclofenac salts. *Int. J. Pharm.* 1999, 187, 163–173.
80. Wachter, W., Buchner, R., and Hefter, G. Hydration of tetraphenylphosphonium and tetraphenylborate ions by dielectric relaxation spectroscopy. *J. Phys. Chem. B.* 2006, 110, 5147–5154.
81. Sipos, P. et al. Raman spectroscopic study of ion-pairing of alkali metal ions with carbonate and sulfate in aqueous solutions *Aust. J. Chem.* 2000, 53, 887–890.
82. Takeda, Y. et al. Extraction of sodium and potassium picrates with 16-crown-5 into various diluents: elucidation of fundamental equilibria determining extraction selectivity for Na^+ over K^+. *Talanta* 1999, 48, 559–569.
83. Harris, M.J., Higuchi, T., and Rytting, J.H. Thermodynamic group contributions from ion-pair extraction equilibriums for use in the prediction of partition coefficients: correlation of surface area with group contributions. *J. Phys. Chem.* 1973, 77, 2694–2703.
84. Gustavii, K. *Acta Pharm. Suecica* Determination of amines and quaternary ammonium ions as complexes with picrate. 1967, 4, 233–246.
85. Modin, R., Schill, G. Quantitative determinations by ion-pair extraction 1: ion-pairs of quaternary ammonium ions with organic anions. *Acta Phram. Sue.* 1967, 4, 301–326.
86. Yeh, K.C. and Higuchi, W.I. Influence of chain length on oil–water ion-pair partitioning behavior of p-alkylpyridinium chlorides. *J. Pharm. Sci.* 1972, 61, 1648–1650.
87. Cecchi, T., Pucciarelli, F., and Passamonti, P. Extended thermodynamic approach to ion interaction chromatography: influence of the organic modifier concentration. *Chromatographia* 2003, 58, 411–419.
88. Treiner, C. and Chattopadhyay, A.K. The Setchenov constant of benzene in non-aqueous electrolyte solutions: alkali–metal halides and aliphatic and aromatic salts in methanol at 298.15 K. *J. Chem Soc. Faraday Trans. 1* 1983, 79, 2915–2927.

89. Marcus, Y. Solvent release upon ion association from entropy data. *Solution Chem.* 1987, 16, 735–744.

90. Beck, M.T. and Nagypal, I. *Chemistry of Complex Equilibria*, Ellis Horwood: Chichester, 1990.

91. Martell, A.E. and Motekaitis, R.J. *The Determination and Use of Stability Constants*, VCH: New York, 1992.

92. Fuoss, R.M. and Hsia, K.L. Association of 1-1 salts in water. *Proc. Natl. Acad. Sci. USA.* 1967, 57, 1500–1557.

93. Fernandez-Prini, R. and Justice, J.C. Evaluation of the solubility of electrolytes from conductivity measurements. *Pure Appl. Chem.* 1984, 56, 541–547.

94. Barthel, J., Wachter, R., and Gores, H.J. In *Modern Aspects of Electrochemistry*, Conway, B.E. et al., Eds., 1979, Vol. 13, p. 1.

95. Hefter, G.T. Calculation of liquid junction potentials for equilibrium studies. *Anal. Chem.* 1982, 54, 2518–2524.

96. Buchner, R., Chen, T., and Hefter, G. Complexity in 'simple' electrolyte solutions: ion-pairing in $MgSO_{4(aq)}$. *J. Phys. Chem. B.* 2004, 108, 2365–2375.

97. Hefter, G.T. When spectroscopy fails: measurement of ion-pairing. *Pure Appl. Chem.* 2006, 78, 1571–1586.

98. Pytkowicz, R.M. In *Activity Coefficients in Electrolyte Solutions*, Pytkowicz, R.M., Ed., CRC Press: Boca Raton, FL, 1991, Vols. I and II.

99. Chen, T., Hefter, G., and Buchner, R. Dielectric spectroscopy of aqueous solutions of KCl and CsCl. *J. Phys. Chem.* 2003, 107, 4025–4031.

100. Barthel, J., Hetzenauer, H., and Buchner, R. Dielectric relaxation of aqueous electrolyte solutions II: ion-pair relaxation of 1:2, 2:1, and 2:2 electrolytes, *Ber. Bunsen Ges. Phys. Chem.* 1992, 96, 1424–1432.

101. Kaatze, U., Hushcha, T.O., and Eggers, F. Ultrasonic broadband spectrometry of liquids: a research tool in pure and applied chemistry and chemical physics. *J. Solution Chem.* 2000, 29, 299–368.

102. Dai, J. and Carr, P.W. Role of ion-pairing in anionic additive effects on the separation of cationic drugs in reversed-phase liquid chromatography. *J. Chromatogr. A.* 2005, 1072, 169–184.

103. Mbuna, J. et al. Capillary zone electrophoretic studies of ion association between inorganic anions and tetraalkylammonium ions in aqueous dioxane media. *J. Chromatogr. A.* 2005, 1069, 261–270.

104. Dai, J. et al. Effect of anionic additive type on ion-pair formation constants of basic pharmaceuticals. *J. Chromatogr. A.* 2005, 1069, 225–234.

105. Steiner, S.A., Watson, D.M., and Fritz, J.S. Ion association with alkylammonium cations for separation of anions by capillary electrophoresis. *J. Chromatogr. A.* 2005, 1085, 170–175.

106. Motomizu, S. and Takayanagi, T. Electrophoretic mobility study of ion–ion interactions in an aqueous solution. *J. Chromatogr. A.* 1999, 853, 63–69.

107. Takayanagi, T., Wada, E., and Motomizu, S. Electrophoretic mobility study of ion association between aromatic anions and quaternary ammonium ions in aqueous solution. *Analyst* 1997, 122, 57–62.

108. Takayanagi, T., Wada, E., and Motomizu, S. Separation of divalent aromatic anions by capillary zone electrophoresis using multipoint ion association with divalent quaternary ammonium ions. *Analyst* 1997, 122, 1387–1392.

109. Takayanagi, T., Tanaka, H., and Motomizu, S. Ion association reaction between divalent anionic azo dyes and hydrophobic quaternary ammonium ions in aqueous solution. *Anal. Sci.* 1997, 13, 11–18.

110. Popa, T.V., Mant, C.T., and Hodges, R.S. Capillary electrophoresis of amphipathic α-helical peptide diastereomers. *Electrophoresis* 2004, 25, 94–107.
111. Popa, T.V., Mant, C.T., and Hodges, R.S. Ion interaction capillary zone electrophoresis of cationic proteomic peptide standards. *J. Chromatogr. A.* 2006, 1111 192–199.
112. Popa, T.V. et al. Capillary zone electrophoresis of α-helical diastereomeric peptide pairs with anionic ion-pairing reagents. *J. Chromatogr. A.* 2004, 1043, 113–122.
113. Mbuna, J. et al. Evaluation of weak ion association between tetraalkylammonium ions and inorganic anions in aqueous solutions by capillary zone electrophoresis. *J. Chromatogr. A.* 2004, 1022, 191–200.
114. Cecchi T. Influence of the chain length of the solute ion: a chromatographic method for the determination of ion-pairing constants. *J. Sep. Sci.* 2005, 28, 549–554.

3 Retention Modeling as Function of Mobile Phase Composition

Raw experimental data would be meaningless without some form of analysis to provide understanding. We need models to obtain meaning from observable fact. A model combines interpretation and a depiction of a phenomenon. The two fundamental types of models are theoretical and empirical. The former follows from known theoretical laws or principles; they convert raw data to knowledge and are predictive in their own right. If sound new data are at variance with theoretical predictions, the theory must be improved to take experimental evidence into account. Conversely, empirical models do not adhere to any theoretical basis; raw data are used to describe the system response.

Ion pairing chromatography (IPC) is often regarded as a mature technique that allows thousand of chromatographers to solve problems successfully in a wide range of applications. However, based on expanded optimization parameter space (pH, type, and lipophilicity of ion-pairing reagent [IPR], organic modifier and IPR concentrations, ionic strength, stationary phase packing) and analyte differences (non-ionic, ionizable, and ionic solutes), the rational selection of optimal experimental conditions that can provide adequate resolution in reasonable run time is a major challenge in the development of a chromatographic method. The success of optimization may become a difficult task in IPC, but the attainment of the maximum performance of separation is vital.

During the infancy of IPC, retention prediction commonly faced trial-and-error procedures that attempted to make the problem univariate, holding all experimental conditions constant except one. This one-at-a-time changing of parameters, without regard to parameter interactions, is still practiced and may, in a time consuming way, improve performance. The description of the dependence of retention on the mobile phase composition parameters is the focus of interest of model makers because an *a priori* retention prediction is highly desirable. Optimization is finding the unique combination of values of adjustable parameters that yields the best performance possible for a set of requirements.

3.1 THEORETICAL MODELS OF ION-PAIR CHROMATOGRAPHY (IPC)

Although the daily grind of processing large sample sets translates chromatography into a practical analytical tool, the theory is not a regression to the earlier days of IPC. This section critically and, as far as possible, chronologically reviews the theoretical schemes developed to describe an IPC system. Stoichiometric models are not described in detail since they are not credited with firm foundations in physical chemistry.

3.1.1 STOICHIOMETRIC MODELS

Stoichiometric models forged the first rationalization of the retention patterns of IPC. All stoichiometric models are pictorial and do not need sophisticated mathematical descriptions of analyte retention. What is the link between ion-pairing and chromatography?

The original concept of ion-pair liquid extraction of ionized solute and partition into an immiscible organic phase was sympathetically adopted by chromatographers. In RP-HPLC separations performed on non-polar stationary phases with polar hydro-organic eluents, polar compounds elute first and ionic solute retention is usually poor because of their high affinity to the polar mobile phase. To achieve the adequate retention pre-requisite to good resolution, the mobile phase was supplemented with a specific IPR, a large organic ion oppositely charged to the analyte of interest [1,2].

These additives were originally supposed to be effective because they form ion-pairs with the analyte in bulk eluent, and for this reason the technique was called ion-pair chromatography [3–5] and soap chromatography because earlier IPRs were tensioactive compounds [6]. An ion-pair acts as a single unit in determining conductivity, kinetic behavior, osmotic properties, and chromatographic retention. For symmetrical ion-pairs, the partners' charges are shielded; the polarity of the duplex is greatly attenuated and its lipophilicity increased, and the RP-HPLC retention of the paired analyte is easily predicted to increase. Improved efficiency and resolution are additional and valuable collateral effects.

However, it was quickly realized during the introduction of this technique that an IPR having a hydrophobic region to interact with the stationary phase was prone to adsorb onto the reversed phase chromatographic bed because (1) the full impact of the reagent was felt after many void volumes of the column were displaced, (2) if the analyte and the IPR carried the same charge status, analyte retention decreased upon introduction of the IPR in the eluent, and (3) when the presence of IPRs in the eluent was discontinued, the previous column retentive behavior was not restored promptly. The IPR adsorption could be measured and modeled via the adsorption isotherm [6].

Model makers named the technique solvent generated ion exchange [7] and hydrophobic chromatography with dynamically coated stationary phase [8], thereby emphasizing a dynamic ion exchange model.

Lipophilic ions first adsorb at the surface of the stationary phase, and the dynamically generated charge sites provide an ion exchange character that explains the retention of oppositely charged analytes [7–11]. This retention mechanism does not explain the contribution of solute hydrophobicity to retention because it should not be relevant if retention is only charge driven. It can be speculated that both mechanisms act and the extent to which one is more significant than the other depends on the experimental set-up and the nature of the IPR [12].

Many subsequent stoichiometric mixed mode models are based on various combinations of these ion-pair and dynamic ion exchange extreme mechanisms. The effect of the IPR counter ion [13] and the reduction of available hydrophobic surfaces

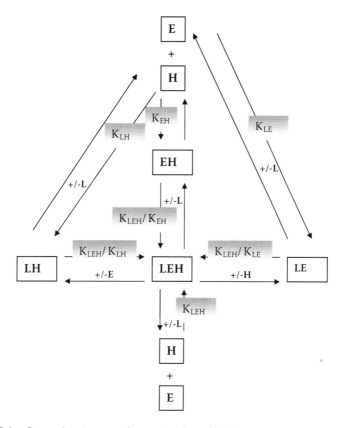

FIGURE 3.1 Comprehensive retention mechanism of IPC.

of the stationary phase [14] were considered to concur to model retention. The most comprehensive survey of all equilibria supposed to occur in a chromatographic system, that is, the retention mechanism upon which the stoichiometric theory is structured, is depicted in Figure 3.1 [1,3]. Since all stoichiometric models suggest the same structure of the adsorbed complex, they yield identical retention equations [1]. In a chromatographic system, the main equilibria are (1) the adsorption of the analyte (E) onto the stationary phase hydrocarbonaceous ligand site (L), (2) the adsorption of IPR (H) onto L, (3) the ion-pair formation in the mobile phase (EH), (4) the ion-pair formation in the stationary phase (LEH), (5) and the displacement of H by E. The equilibrium constants for these equilibria are, respectively, K_{LE}, K_{LH}, K_{EH}, K_{LEH}, and K_{LE}/K_{LH}. These models are termed "stoichiometric" because they use stoichiometric equilibrium constants instead of thermodynamic constants (see below) to describe the ion association process.

The mechanism of IPC was hotly debated for years and continues as a matter of debate. The retention model of Bidlingmeyer [15,16] is more comprehensive than the ion-pair and dynamic ion exchange models. Lipophilic IPRs, due to their adsorbophilic nature, dynamically adsorb onto the alkyl-bonded apolar surface of the stationary phase, forming a primary charged ion layer and, together with counter ions in the diffuse outer region, develop an electrical double layer. The retention of the sample results from both electrical and van der Waals forces by means of a mixed retention mechanism. Selectivity is determined by a number of variables such as eluent ionic strength, pH and composition, and IPR concentration and lipophilicity that may be tuned easily to achieve tailor-made separations with adequate retention for both ionized and non-ionized analytes in a single run.

This ion interaction retention model of IPC emphasized the role played by the electrical double layer in enhancing analyte retention even if retention modeling was only qualitatively attempted. It was soon realized that the analyte transfer through an electrified interface could not be properly described without dealing with electrochemical potentials. An important drawback shared by all stoichiometric models was neglecting the establishment of the stationary phase electrostatic potential. It is important to note that not even the most recent stoichiometric comprehensive models for both classical [17] and neoteric [18] IPRs can give a true description of the retention mechanism because stoichiometric constants are not actually constant in the presence of a stationary phase–bulk eluent electrified interface [19,20]. These observations led to the development of non-stoichiometric models of IPC. Since stoichiometric models are not well founded in physical chemistry, in the interest of brevity they will not be described in more depth.

3.1.2 NON-STOICHIOMETRIC MODELS

All non-stoichiometric models adopt the electrical double layer concept and disagree with the stoichiometric hypothesis of an electroneutral stationary phase; they emphasize the higher adsorbophilicity of the IPR compared to that of its counter ion: a surface excess of IPR ions generates a primary charged layer and a charged interface. Like-sign co-ions are repelled from the surface while IPR counter ions are attracted by the charged surface.

There are two versions of the physical description of this system. According to the Gouy-Chapman (G-C) theory, counter ion thermal energy runs counter to the electrostatic attraction and a secondary diffuse layer in which the potential decays almost exponentially because of screening effects is generated. Both layers are under dynamic equilibrium. The electrical potential difference, Ψ°, between the stationary phase and the bulk eluent can be theoretically estimated. Figure 3.2 depicts the G-C model.

Conversely, according to the description of the electrical double layer based on the Stern-Gouy-Chapman (S-G-C) version of the theory [24], counter ions cannot get closer to the surface than a certain distance (plane of closest approach of counter ions). Chemically adsorbed ions are located at the inner Helmholtz plane (IHP), while non-chemically adsorbed ions are located in the outer Helmholtz plane (OHP) at a distance x_o from the surface. The potential difference between this plane and the bulk solution is Ψ_{OHP}. In this version of the theory, Ψ_{OHP} replaces Ψ° in all equations. Two regions are discernible in the double layer: the compact area between the charged surface and the OHP in which the potential decays linearly and the diffuse layer in which the potential decay is almost exponential due to screening effects.

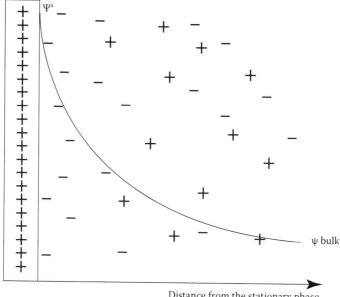

Distance from the stationary phase

FIGURE 3.2 Gouy-Chapman double layer model.

The higher the ionic strength, the faster the potential decays. Figure 3.3 illustrates the S-G-C model.

The first attempts of Bidlingmeyer and co-workers [15,16] to formulate an ion interaction model quantitatively [21–23] did not provide a rigorous description of the system. Stranahan and Deming [22] accounted for electrostatic effects via a simplified activity coefficient in the stationary phase. An interfacial tension decrease with increasing IPR concentration was considered responsible for the appearance of maxima in the plot of retention factor, k versus IPR concentration, but experimental results were at odds with known surfactant chemistry.

Two genuine electrostatic non-stoichiometric theories were developed: one by Ståhlberg and co-workers and the other by Cantwell and co-workers. They considered the processes involved in IPC as based solely on the formation of the electrical double layer and disregarded the ion-pairing process in the bulk eluent.

In the electrostatic model of Ståhlberg [19], the description of the double layer is based on the G-C theory and solute retention is governed by the analyte free energy of adsorption $\Delta G^{\circ}_{t,E}$ that comprises two components: the chemical part known as $\Delta G^{\circ}_{c,E}$ parallels the analyte hydrophobicity and corresponds to the free energy of adsorption of the analyte in the absence of the IPR; it is assumed to remain constant also in the presence of the IPR. The second component known as $\Delta G^{\circ}_{el,E}$ represents the electrostatic work involved in the transfer of a charged analyte from the bulk eluent to the charged stationary phase. Since $\Delta G^{\circ}_{el,E}$ is related to the surface electrostatic potential, Ψ°, according the following relationship:

$$\Delta G^{\circ}_{el,E} = z_E F \, \Psi^{\circ}$$

(3.1)

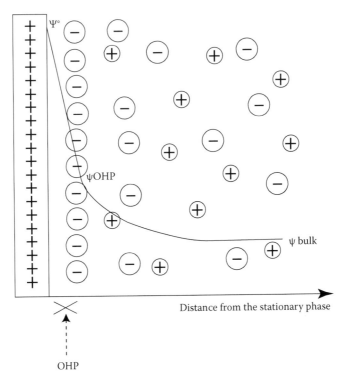

FIGURE 3.3   Stern-Gouy-Chapman model, in which only the OHP is indicated.

where z_E is the charge of the analyte and F is the Faraday constant; it follows that

$$\Delta G^{\circ}_{t,E} = \Delta G^{\circ}_{c,E} + z_E F\, \Psi^{\circ} \qquad (3.2)$$

and the thermodynamic equilibrium constant ($K_{t,E}$) for the adsorption of the analyte E onto the stationary phase is

$$K_{t,E} = \exp\left(-\Delta G^{\circ}_{t,E}/RT\right) \qquad (3.3)$$

where R is the gas constant, and T is the absolute temperature. Since the model does not take into account the possibility from basic chromatographic theory that the analyte adsorbs as an ion-pair, the following relationships can be easily obtained if (1) the activities are assumed to tend to the concentrations and (2) the available ligand sites are assumed to be constant even in the presence of adsorbed IPR:

$$k = \phi \frac{[LE]}{[E]} = \phi K_{t,E} = \phi \exp\left(-\frac{\Delta G^{\circ}_{c,E}}{RT} - \frac{z_E F\Psi^{\circ}}{RT}\right) = k_o \exp\left(-\frac{z_E F\Psi^{\circ}}{RT}\right) \qquad (3.4)$$

where ϕ is the ratio of the surface area of the column to the mobile phase volume (inverse phase ratio of the column), $[LE]$, and $[E]$, are, respectively, the surface and mobile phase concentrations of the analyte, k_0 is the retention factor of the analyte when the surface potential is zero, that is, in the absence of the IPR. To obtain the rigorous relationship for $\Psi°$, the porous nature of the particles packed in a column would require the solving of the Poisson-Boltzmann equation in cylindrical coordinates. Usually, if the ionic strength is high enough [24], the inverse Debye length is low and planar surface geometry can be used, thereby obtaining the following rigorous relationship [25]:

$$\Psi° = \frac{2RT}{F} \ln \left\{ \frac{[LH] \cdot |z_H| F}{\left(8\varepsilon_0 \varepsilon_r RT \sum_i c_{0i} \right)^{\frac{1}{2}}} + \left[\frac{([LH] \cdot z_H F)^2}{8\varepsilon_0 \varepsilon_r RT \sum_i c_{0i}} + 1 \right]^{\frac{1}{2}} \right\} \tag{3.5}$$

where z_H is the charge of the IPR, $[LH]$ is the surface concentration of the IPR, ε_0 is the electrical permittivity of the vacuum, ε_r is the dielectric constant of the mobile phase, and Σc_{0i} is the mobile phase concentration (mM) of the electrolyte ions that are assumed to be singly charged. $\Psi°$ is positive or negative, according to the charge status of the IPR. Ståhlberg et al. did not use the rigorous Equation 3.5; they used a simpler linearized expression:

$$\Psi° = \frac{[LH] \cdot z_H F}{\kappa \varepsilon_0 \varepsilon_r} \tag{3.6}$$

(where κ is the inverse Debye length) even if the threshold limit (25 mV) above which the simplified expression does not hold can be easily exceeded under usual chromatographic practice [24]. To calculate the surface potential, $[LH]$ is needed and depends on the mobile phase concentration of the IPR $[H]$ according to its adsorption isotherm. The potential modified Langmuir adsorption isotherm correctly describes the IPR adsorption but to avoid a too-complex expression, the adsorption isotherm was linearized, thereby introducing a further approximation in the retention model. Unfortunately, since many IPRs are very adsorbophilic, $[LH]$ can be high even at very low $[H]$ and the linearization is not usually acceptable under typical experimental conditions in IPC [20,24,26,27]. A method that helps model makers determine whether the linearization of the potential modified Langmuir isotherm is feasible, and accordingly, whether simplified retention equations can be properly used was developed recently [27]. In the series of approximations devised to obtain an expression for practical testing, a surface potential empirically obtained from retention data was also used even if different apparent surface potentials were obtained when different probe solutes were tested [26]; a series expansion of a double logarithmic term was also necessary [19]. Note, however, that the major concern that arises upon reviewing this non-stoichiometric retention model is not the series of approximations or the empirical "contamination" of the theoretical model, but just the suggested mechanism that fails to account for the formation of ion-pairs. We will show that this retention model ignores pairing equilibria and thus is not able to explain well-built experimental evidence, even if a multi-site occupancy model [28] is considered.

Cantwell and co-workers submitted the second genuine electrostatic model; the theory is reviewed in Reference 29 and described as a surface adsorption, diffuse layer ion exchange double layer model. The description of the electrical double layer adopted the Stern-Gouy-Chapman (SGC) version of the theory [30]. The role of the diffuse part of the double layer in enhancing retention was emphasized by assigning a stoichiometric constant for the exchange of the solute ion between the bulk of the mobile phase and the diffuse layer. However, the impact of the diffuse layer on organic ion retention was demonstrated to be residual [19].

To use the theory for retention data modeling [31–33], a family of adsorption isotherms of the IPR (each isotherm is measured at constant eluent ionic strength) is needed; hence testing and employment of the model are not straightforward. Both genuine electrostatic retention models share the view that the "chemical" part of the free energy can be considered constant after the addition of the IPR, but the theory by Cantwell and co-workers is more complex, from both conceptual and practical views. Like the model by Ståhlberg and co-workers, it neglects the formation of ion-pairs and hence the same major criticism applies to both theories.

Clearly, even if all retention models have had their disciples and were experimentally confirmed, a plain fit of experimental results does not ensure that a theory is correct. A chromatographic theory has two major goals: (1) to explain raw data and describe experimental evidence and (2) to comply with fundamental physical chemistry.

If the crucial drawback of stoichiometric models is neglect of the demonstrated development of the electrical double layer, a serious setback of electrostatic models is ignoring the experimental proof [34–48] of the formation of chemical complexes between oppositely charged analyte and IPR, discussed in detail in Chapter 2.5.3. In a landmark paper related to theoretical modeling of ion-pairing, Popa and co-workers intensively studied ion-pairing in the CE separation of diastereomeric peptide pairs [42]. Despite the absence of a hydrophobic surface, a clear separation according to the analyte hydrophobicity was obtained. Since separation improves with increasing IPR concentration and hydrophobicity, the CE method demonstrates ion-pairing in solution [43] and is an eligible technique to estimate ion-pairing equilibrium constants (see Chapter 2.7.6). All these experimental proofs [34–48] definitively impair and bias the theoretical retention mechanisms of the purely electrostatic models of Ståhlberg et al. and of Cantwell et al.

Lu and co-workers faced the modeling of IPC retention [49] via a statistical thermodynamic method combined with the Freundlich adsorption isotherm to obtain a retention equation as a function of IPC concentration, ionic strength, and organic modifier percentage in the eluent. They accounted for both electrostatic and molecular pairing interactions and obtained the chemical potentials of different species even if the description of these interactions at a molecular level was very complicated. For example, the chemical potential of the IPR in the adsorbed phase depends on (1) the internal partition functions of the solvents and of the IPR and (2) on the potential energy of the eluent and of the IPR on the stationary phase; the latter, in turn, depends, among other factors, on constants correlated to the molecular size of both the IPR and of each solvent and the extent of the molecular interaction of the IPR and each solvent. For the analyte, the expression for the chemical potential is even more complicated. Lu's retention equations fitted experimental results but the estimates of the copious fitting parameters (whose physical meaning is clear even if complicated) are impossible to comment upon because many of the cited parameters are not known. The author speculated that further investigation of the coefficient of the fitted equations was necessary.

A simpler description of the system is desired and it can be obtained using equilibrium constants that describe all equilibria in the system; obviously only a thermodynamic equilibrium constant must be made use of since, as discussed above, at variance with stoichiometric constants, they account for all the molecular level interactions in a very simple way, and notably, their chromatographic and non-chromatographic estimates can be compared to validate the retention mechanism they describe.

Starting from this idea, Cecchi and co-workers submitted an extended thermodynamic theoretical treatment of the retention behavior that covers and comprehends both stoichiometric and genuine electrostatic models but surpasses them [20,26,27,50–64]. The subject is not difficult and a tutorial description is given below. More detailed and comprehensive descriptions of the model can be found elsewhere [20,63].

Basically, the retention model capitalized on the retention mechanism of the most complete stoichiometric model [1,3] detailed in Figure 3.1, but it describes the ionic duplex formation at a thermodynamic (and not stoichiometric) level. It also accounts for the electrical double layer that results from IPR adsorption. Principally, to describe analyte retention as a function of the mobile phase composition we need the surface coverage of the IPR to calculate the surface potential, $\Psi°$, according to the rigorous Gouy-Chapman equation (3.5), even if for high [51,53,55] or low [52,56] potential, the rigorous expression can be approximated (see Equation 3.6 above). The surface concentration of the IPR is obtained via the Freundlich adsorption equation, related to the potential modified Langmuir adsorption isotherm, that holds for the adsorption of hydrophobic ions such as IPRs [65]. The unambiguous limit between theoretical and empirical uses of the Freundlich isotherm was recently ascertained [27]. The Freundlich relationship is:

$$[LH] = a[H]^b \tag{3.7}$$

where a and b are constants that can be experimentally evaluated.

The thermodynamic equilibrium constants of all equilibria in the chromatographic system can be derived from basic thermodynamic considerations. When the charge status of the analyte and IIR are the same, or if the analyte is neutral, ion-pairing equilibria do not apply. For the condition of equilibrium, it holds that the electrochemical potentials (μ) of species in the stationary and mobile phases are equal:

$$\mu_{LE} = \mu_L + \mu_E \tag{3.8}$$

$$\mu_{LH} = \mu_L + \mu_H \tag{3.9}$$

$$\mu_{LEH} = \mu_L + \mu_E + \mu_H \tag{3.10}$$

$$\mu_{EH} = \mu_E + \mu_H \tag{3.11}$$

where μ represents the electrochemical potential for each species. If the chemical species is charged and it is present in a phase whose electrostatic potential is not zero, the electrochemical potential takes into account the electrostatic energy of the interaction of the charged species with the potential, for example:

$$\mu_{LH} = \mu_{LH}^0 + RT \ln a_{LH} + z_H F \Psi^\circ \qquad (3.12)$$

where μ^0 represents the electrochemical potential for the standard state and a_{LH} is the activity of the adsorbed H. Since Ψ° is of the same sign as z_H, it always runs counter to further adsorption of the IPR because it increases its electrochemical potential. Similar expressions can be obtained for the electrostatic potentials of LE: in this case, if E is oppositely charged to the IPR, Ψ° favors its adsorption onto the stationary phase. If it is in the same charge status as the IPR, it experiences a repulsive interaction with the stationary phase and a decrease in its retention (not easily predictable by the ion-pair model) is obtained. Conversely, for neutral chemical species (L, EHL, EH) or charged chemical species (E, H) in the electrically neutral bulk mobile phase, we have expressions similar to the following

$$\mu_L = \mu_L^0 + RT \ln a_L \qquad (3.13)$$

Since a thermodynamic equilibrium constant has the general expression:

$$K = \exp(-\Delta\mu^0/RT) \qquad (3.14)$$

the following relationships are easily obtained for each thermodynamic equilibrium constant (a represents the activity of each species):

$$K_{LE} = \frac{a_{LE}}{a_L \cdot a_E} \exp\left(\frac{z_E F \Psi^\circ}{RT}\right) \qquad (3.15)$$

$$K_{LH} = \frac{a_{LH}}{a_L \cdot a_H} \exp\left(\frac{z_H F \Psi^\circ}{RT}\right) \qquad (3.16)$$

$$K_{EH} = \frac{a_{EH}}{a_E \cdot a_H} \qquad (3.17)$$

$$K_{LEH} = \frac{a_{LEH}}{a_L \cdot a_E \cdot a_H} \qquad (3.18)$$

The thermodynamic equilibrium constants shown by Equations 3.15 and 3.16 match the stoichiometric (or concentration based-) constants of stoichiometric models (see Equations 1 and 3 of Reference 1). Since the latter neglect the modulation of the adsorption of a charged species by the surface potential, they are not constant [19] after the addition of the IPR in the mobile phase. Stoichiometric relationships [19] represent only the ratio of equilibrium concentrations and cannot describe equilibrium in the presence of electrostatic interactions. In their stoichiometric approach,

Knox and Hartwick [1] included counter ions to ensure electrical neutrality in the adsorption process. We submit that their approach is not acceptable on the basis of the extensive studies that demonstrate the development of surface potential as a consequence of the stronger adsorbophilic attitudes of the IPR compared to that of its counter ion [31]. From basic chromatographic theory, if ion-pairing equilibria are taken into account, the retention factor of the analyte, k, is given by:

$$k = \phi \frac{[LE]+[LEH]}{[E]+[EH]} \tag{3.19}$$

where $[LEH]$ and $[EH]$ are, respectively, the surface and mobile phase concentration of the complex EH, and $[LE]$ and $[E]$ are, respectively, the surface and mobile phase concentration of E. Since the coverage of the stationary phase accessible sites is expected to be small under trace conditions and the total ligand concentration $[L]_T$ is conserved, it can be written [1,3] that:

$$[L]_T = [L] + [LH] \tag{3.20}$$

where $[L]$ is the surface concentration of free adsorption sites.

When the analyte mobile phase concentration is small, only a negligible fraction of the IIR is in the form of a complex, hence its concentration $[H]$ in the eluent can be considered invariant [3]. Both the pairing ion isotherm and the surface potential are unchanged by the presence of the sample ion [31,33]. In this case [20], analyte retention as a function of the mobile and stationary phase concentrations of the IIR can be described, respectively, by the following expressions:

$$k = \phi[L]_T \frac{K_{LE}\dfrac{\gamma_L\gamma_E}{\gamma_{LE}}epx(-z_E F\Psi^\circ/RT)+K_{LEH}\dfrac{\gamma_L\gamma_E\gamma_H}{\gamma_{LEH}}[H]}{\left(1+K_{EH}\dfrac{\gamma_E\gamma_H}{\gamma_{EH}}[H]\right)\left(1+K_{LH}\dfrac{\gamma_L\gamma_H}{\gamma_{LH}}[H]epx(-z_H F\Psi^\circ/RT)\right)} \tag{3.21}$$

$$k = \phi([L]_T-[LH])\frac{K_{LE}\dfrac{\gamma_L\gamma_E}{\gamma_{LE}}epx(-z_E F\Psi^\circ/RT)+K_{LEH}\dfrac{\gamma_L\gamma_E\gamma_H}{\gamma_{LEH}}\dfrac{[LH]^{1/b}}{a/b}}{\left(1+K_{EH}\dfrac{\gamma_E\gamma_H}{\gamma_{EH}}\dfrac{[LH]^{1/b}}{a/b}\right)} \tag{3.22}$$

The modulation of analyte retention by the electrostatic surface potential is described by the first term in the numerator of Equation 3.21. If the analyte is oppositely (similarly) charged to the IPR, it experiences an attractive (repulsive) interaction with the stationary phase; hence its retention would monotonously increase (decrease) with increasing IPR concentration in the absence of other interactions. The second term in the numerator of Equation 3.21 is related to ion-pair formation at the stationary phase that may ensue from four interdependent equilibria: (1) the simultaneous adsorption of both E and H onto L, (2) the adsorption of E onto LH, (3) the adsorption of H onto LE, and (4) the adsorption of EH onto L.

Ion pairing at the stationary phase always results in a retention increase, while ion-pair formation in the mobile phase, described by the left factor of the denominator of Equation 3.21, decreases retention because the analyte is moved from the stationary phase toward the eluent. Both terms concerning ion-pair formation are missing if the analyte and IIR are similarly charged, or if the analyte is neutral.

The right factor of the denominator of Equation 3.21 describes the competition between the analyte and the IPR for the available ligand sites. Since they decrease with increasing IPR concentration, retention is expected to decrease at high surface coverage, especially for analytes that are neutral or have the same charge status as the IPR. Conversely, analytes oppositely charged to the IPR, would be electrostatically attracted by a very crowded surface and multilayer formation is easily predicted to prevent surface exclusion phenomena. In Chapter 10 it will be demonstrated that Equation 3.21 is algebraically equivalent to Equation 10.12. Equation 3.22 parallels Equation 3.21 except adsorption competitions are accounted for by factor $([L]_T - [LH])$.

When the exact G-C relationship for the potential (Equation 3.5) and the Freundlich equation for the adsorption isotherm of the IPR (Equation 3.7) are substituted in Equations 3.21 and 3.22, the following expressions are respectively obtained [20]:

$$k = \frac{c_1\left\{a[H]^b f + [(a[H]^b f)^2 + 1]^{\frac{1}{2}}\right\}^{\pm 2|z_E|} + c_2[H]}{(1+c_3[H])\left\{1+c_4[H]\left\{a[H]^b f + [(a[H]^b f)^2 + 1]^{\frac{1}{2}}\right\}^{-2|z_H|}\right\}} \tag{3.23}$$

$$k = (d_4 - [LH])\frac{d_1\left\{[LH]f + [([LH]f)^2 + 1]^{\frac{1}{2}}\right\}^{\pm 2|z_E|} + d_2[LH]^{1/b}}{(1+d_3[LH]^{1/b})} \tag{3.24}$$

where f (m^2/mol) is a constant if the ionic strength is kept constant, which can be evaluated from the eluent composition and temperature:

$$f = \frac{|z_H| F}{8\varepsilon_0\varepsilon_r RT \sum_i c_{0i}} \tag{3.25}$$

In the exponent of the first term of the numerator of Equation 3.23, the plus (minus) sign applies for oppositely (similarly) charged analytes and IPR, since the surface potential and analyte charge are of opposite (like) sign and the electrostatic interaction is attractive (repulsive). The ion-pairing terms (c_2 and c_3) are absent for similarly charged analytes and IPR. Equation 3.23 can take the magnitude of the analyte charge z_E into consideration and correctly predicts that the electrostatic interaction is stronger for multiply charged analytes [58]. When Equations 3.23 and 3.24 are fitted to experimental data, excellent results are obtained [20,26,27,50–64]. Parameters c_1-c_4 have a clear physical meaning [20] and this allows the parameter estimates to be commented upon:

$$c_1 = \phi[L]_T K_{LE} \frac{\gamma_L \gamma_E}{\gamma_{LE}} \tag{3.26}$$

$$c_2 = \phi[L]_T K_{LEH} \frac{\gamma_L \gamma_E \gamma_H}{\gamma_{LEH}} \tag{3.27}$$

$$c_3 = K_{EH} \frac{\gamma_E \gamma_H}{\gamma_{EH}} \tag{3.28}$$

$$c_4 = K_{LH} \frac{\gamma_L \gamma_H}{\gamma_{LH}} \tag{3.29}$$

Since c_1 represents k_0 (analyte retention factor in the absence of the IPR), it can be obtained from experimental results; hence it was not considered an optimization parameter. However, when c_1 was estimated by fitting of experimental results, the percent error was very low. c_2 is related to the thermodynamic equilibrium constant for ion-pairing at the stationary phase; c_3 represents the equilibrium constant for ion-pairing in the eluent; and c_4, represents the equilibrium constant for IIR adsorption.

The estimates of these parameters are very reliable since they agree with chromatographic and non-chromatographic assessments [20,26,50,58,59,65–69]. An independent proof of the fact that ion-pairing in the eluent occurs is the ability of non-UV-absorbing ions to be detected spectrophotometrically when highly absorbing IPRs are used [67]. This lends strong support to the model and entitles IPC to be considered a valid additional quantitative tool to estimate ion-pairing constants (see Section 2.7 in Chapter 2 to learn about techniques for studying ion-pairing).

The present model can explain experimental evidence that cannot be rationalized by electrostatic retention models. A fixed surface concentration of different IPRs is obtained at a specific eluent concentration of the IPR, according to each IPR-specific adsorption isotherm. It follows that at fixed surface concentration of the IPR, different IPR chain lengths or structures result in (1) a different withdrawal of the IPR toward the mobile phase to form ion-pairs and (2) different pairing equilibrium constants. Hence non-coincidental curves are easily predicted and verified [1] when k is plotted against $[LH]$. Conversely, a genuine electrostatic model [19] for fixed surface concentrations of different IPRs would erroneously predict identical solute retention because the surface potential is the same. If ion-pairing interactions are missing because the analyte and IPR are similarly charged, these curves become coincidental and the electrostatic retention model correctly predicts retention.

The retention model by Cecchi and co-workers also quantitatively faced the prediction of the retention behavior of neutral and zwitterionic analytes in IPC. According to the electrostatic models, at odds with clear experimental data [1,50,52,53], the retention of a neutral solute is not dependent on the presence and concentration of a charged IPR in a chromatographic system. Equation 3.23 is very comprehensive: if z_E is zero [50], it simplifies since ion-pairing does not occur ($c_2 = c_3 = 0$). Adsorption competition models the retention patterns of neutral analytes in IPC and the slight retention decreases of neutral analytes with increasing IIR concentration may be quantitatively explained [50,53].

As far as zwitterionic solutes [54–56,61–62] are concerned, the model predicts that their electrical dipole in the non-homogeneous electrical field that develops at the interphase experiences a torque moment that arranges it parallel to the lines of the field, with the head oppositely charged to the electrostatic potential facing the stationary phase surface. Hence, the electrical interaction is always attractive and it shoves the dipole toward the interphase, where the field is stronger.

This description is similar to considering the electrical dipole equivalent to a fractional electrical charge. The force acting on the dipole or on the equivalent fractional charge was calculated and used to find the electrostatic term in the electrochemical potential of the solute from which the thermodynamic equilibrium constant for its adsorption can be calculated. Both the dipole and the fractional charge mechanisms proved successful in describing the retention patterns of zwitterions [54–56,61,62]. They give the same retention equation, even if the physical meaning of one fitting parameter is different.

$$k = \frac{c_1 \left\{ a[H]^b f + [(a[H]^b f)^2 + 1]^{\frac{1}{2}} \right\}^{2c_2}}{\left\lfloor 1 + c_4 [H] \left\{ a[H]^b f + [(a[H]^b f)^2 + 1]^{\frac{1}{2}} \right\}^{-2} \right\rfloor} \quad (3.30)$$

In Equation 3.30, c_1 and c_4 can be obtained, respectively, from Equations 3.26 and 3.29 and c_2 is related to the molecular dipole [54], or alternatively to the fractional charge [61]: the higher it is, the stronger the retention increase upon IIR addition. Equation 3.30 can be valuable in IPC of life science samples that require keeping peptides at their isoelectric points (zwitterions) to avoid denaturation.

The retention model by Cecchi and co-workers can also treat the simultaneous influences of both the organic modifier and IPR concentrations in the eluent from a bivariate view [57]. The retention equation correctly predicts that retention diminishes with increasing organic modifier concentration and increases with increasing IPR concentration, as expected. Interestingly, this increase is lower at high organic modifier eluent concentration since the organic modifier reduces analyte retention both directly (it decreases the solute free energy of adsorption) and indirectly (it lessens the IPR free energy of adsorption on the stationary phase; hence the electrostatic potential is lower and this results in a retention fall). In the absence of the IPR, the bivariate retention equation reduces to the expression that describes the influence of the organic modifier on RP-HPLC retention [57,70] and this is epistemologically rewarding.

Section 2.5.3 in Chapter 2 expounded upon the hydrophobic ion-pair concept. The peculiarities of this association mode, not even likely in the Bjerrum's model, were elucidated. Electrostatic attraction is only part of the story and solvophobic interactions are crucial to rationalize experimental evidence that often runs counter to the pristine electrostatic description of the process.

The quantitative modeling of the influence of the percentage of the organic modifier in the eluent [57] and the influence of the analyte chain length (lipophilicity) [60] are instructive because they both confirm that hydrophobic ion-pairing is characterized by the opposite needs of electrostatic Bjerrum-type ion-pairing since (1) it amplifies is the sizes of the ions increase because they are forced together by their mutual affinity and (2) it decreases with the decreasing dielectric constant of the medium because the ability of water to force poorly hydrated ions together to decrease their modification of the water structure decreases with increasing percentage of the organic modifier. The estimated values of the ion-pairing equilibrium constant parallel the increasing enforcement of the hydrogen bonded structure of water [47] or the increasing mutual affinity of the organic chains of the ions. This affinity is better described by a stacking free energy, E_{st}, that can be considered a kind of adsorption energy linearly dependent on the concentration of the organic modifier [19, 20]. From the thermodynamic ion-pairing equilibrium constants, the Gibbs free energy of the association can be estimated:

$$\Delta G^\circ = -RT \ln K_{EH} \qquad (3.31)$$

Results by Cecchi and co-workers quantitatively indicated that ion-pairing is characterized by a negative ΔG° of association (spontaneous process); its order of magnitude is about -10 kJ/mol, in agreement with other estimates. However ΔG° increases linearly [60] with increasing organic modifier concentration in the eluent, thereby demonstrating that organic solvents favor ion-pairing via a decrease of the medium dielectric constant, they also run counter to ion-pairing if the solvation is strong; the latter effect predominates with strongly hydrophobic ions.

Interestingly, the estimates of pairing constants obtained by the fitting of retention equations to experimental data increase with increasing analyte chain length (lipophilicity) [60] thereby supporting the hydrophobic ion-pairing concept.

The retention model by Cecchi and co-workers also treats the simultaneous influences of both ionic strength and IPR concentration in the eluent via both monovariate and bivariate methods [59]. To optimize an IPC separation, it is common to increase the IPR concentration to find the best compromise of resolution and chromatographic time. If a compensatory electrolyte is not added to the eluent, the ionic strength increases only as a result of increasing IPR concentration. At high ionic strength (1) the thickness of the diffuse layer is compressed, since more counter ions are present in the eluent (see Figure 3.27), (2) larger adsorption of the IPR is favored because counter ions lower the self-repulsion forces among similarly charged adsorbed IPR ions (Donnan effect) [32,33,66,71]; and (3) electrolytes better shield the surface charge density.

The magnitude of Ψ° is sanctioned by Equation 3.5: its increase due to increased IPR adsorption is overwhelmed by its decrease due to surface charge shielding and a net attenuation of Ψ° is obtained. It follows that f in Equations 3.23 and 3.24 cannot be considered constant and its dependence on ionic concentration according to Equation 3.25 must be explicitly addressed in Equations 3.21 and 3.22 [59]. Moreover, the IPR adsorption isotherm a and b parameters must be obtained under identical experimental conditions, that is, with changing ionic strength [59].

The model additionally considers that solutes may undergo salting in or salting out. According to the solvophobic theory [3]; the former (latter) is predicted when salt addition leads to decreased (increased) surface tension of the eluent. Hydrophobic ions such as IPRs are usually tensioactive salting-in agents [3]. The solvophobic theory shows the dependence of equilibrium constants on ionic strength. When Equations 3.21 and 3.22 are modified based on the above issues, the relationship that can be obtained is algebraically complex but usually may be simplified because salting effects are negligible if the ionic strength increase is not very strong [59].

When both the swamping electrolyte and the IPR concentrations are varied, the retention is correctly described by a bivariate expression [59]. The relationship predicts that the capacity factor of a solute oppositely (similarly) charged to the IPR decreases (increases) with increasing ionic strength, because of the lower net electrostatic attractive (repulsive) potential. The predictions are experimentally verified [59]. Again, it is epistemologically interesting to observe that the retention equation, in the absence of the IPR in the eluent, reduces to that developed and tested in RP-HPLC [72,73].

The model was recently tested to determine whether it was able to model analyte retention in the presence of novel and unusual IPRs (see Chapter 7) such as chaotropic salts and ionic liquids. Chaotropes that break the water structure around them and lipophilic ions (classical IPRs and also ionic liquids) that produce cages around their alkyl chains, thereby disturbing the ordinary water structure, are both inclined to hydrophobic ion-pairing since both are scarcely hydrated. This explains the success of the theory, that is predictive in its own right, when neoteric IPRs are used [64]. Recently a stoichiometric model (*vide supra*) was put forward to describe retention of analytes in the presence of chaotropic IPRs in eluents [18] but its description of the system is not adequate [64].

It was recently clarified [74] that retention of polar solutes is better described by adsorption at the bonded phase–solvent interface, rather than by a partitioning process in which the analytes fully embed themselves into the bonded phase. This corroborates the importance given by the present model to the interfacial region and surface concentration.

It is worth noting that thermodynamic treatment of chromatographic equilibria does not contrast with the fact that, including in many other separative systems chromatographic processes, diffusion and migration are two essential physical chemistry phenomena that belong to the category of linear non-equilibrium thermodynamics [75]. The assumption of local equilibrium in non-equilibrium thermodynamics is a foundation stone [75] that ensures that we can use the thermodynamic equilibrium properties of the system to describe it. In conclusion, the claims for a cut-above theory, epistemologically speaking, are sustained by the following arguments:

- The model is inclusive: the retention equation quantitatively predicts the behavior of charged, multiply charged, neutral, and zwitterionic solutes in the presence of both classical and neoteric IPRs, and can also quantitatively consider the influences of organic modifier concentration and ionic strength.
- The model, in the absence of IPR, from an algebraic view, demonstrates the well known relationships in RP-HPLC of the influence of the organic modifier or the ionic strength of the eluent on analyte retention.
- The theory can comprehend the laws sanctioned by previous theories. First, if the influence of surface potential is disallowed, Equation 3.21 reduces to Equation (2) of the stoichiometric model of Knox and Hartwick [1] except

that their equilibria consider the counter ions necessary to ensure the electrical neutrality of the stationary phase, in open conflict with the experimental proof of the development of surface potential. Second, if the pairing equilibria are neglected, Equation 3.21 reduces to the expression sanctioned by the pure electrostatic approach (see Equation (26) of Reference 19). I adsorption competitions are also negligible, Equation (4a) of Reference 19 may be obtained. This model improves on previous milestone retention models.

- The theory is evidence based; it can explain: (1) the presence of different theoretical curves when k is plotted as a function of the stationary phase concentrations for different IPRs; (2) the influence of IPR concentration on the ratio of the retention of two different analytes; and (3) the influence of analyte nature on the k/k_0 ratio if the experimental conditions are the same. These experimental behaviors cannot be explained by other genuine electrostatic retention models because they arise from complex formation. The present model was the first to take into account at a thermodynamic level these recently definitively demonstrated equilibria in the mobile phase.

- The theory can quantitatively predict that the lower the likelihood of ion-pairing interactions, the better the genuine electrostatic retention model works: (1) at high methanol percentages in the eluent (retention is basically electrostatic since hydrophobic ion-pairing is strongly attenuated); (2) if the chain length of the analyte is shorter (the solute structure approaches the point charge description typical of the electrostatic theories); and (3) for similarly charged analytes and IPR because ion-pairing interactions obviously do not apply to them [60].

- Fitting parameters have an unambiguous physical meaning and their estimates are reliable since they compare well with literature chromatographic and non-chromatographic estimates.

As a concluding remark, we must comment on the name commonly given to this separation strategy on the basis of the most plausible retention mechanism. Since it was confirmed that the simple formation of ion-pairs in an eluent does not increase but decreases analyte retention (because the analyte is withdrawn from the stationary phase into the mobile phase) [1,20], the "ion-pair chromatography" phrase does not seem to be the best.

Pioneering work by Bidlingmeyer and co-workers explained the essence of this separation mode [15,16] that is completely different from simple stoichiometric ion-pairing. They qualitatively demonstrated that the retention increase of a charged analyte follows from a complex ion interaction mechanism due to the dynamic adsorption of the IPR hydrophobic ions onto the reversed phase of the stationary phase surface. As a result, "ion interaction chromatography" is a more fitting description. Nevertheless, the phrase "ion-pair chromatography" was commonly used also by model makers whose retention models do not postulate the formation of ion-pairs [19,29]. The extended physical–chemical treatment by Cecchi and co-workers [20,26,27,50–64] quantitatively confirms that the *ion interaction chromatography* term is certainly the most correct one. Nevertheless, in this book, the common *IPC* term will be used in the interest of clarity. Figure 3.4. summarizes the influences of various parameters on the retention of a model analyte and quantitatively explains the origin of retention maxima.

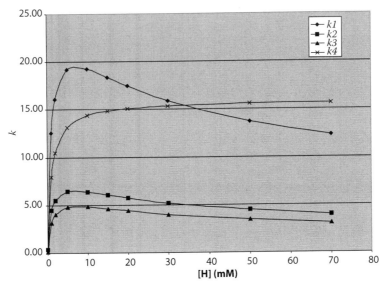

FIGURE 3.4 Retention of model analyte as function of IPR eluent concentration. $k1$: $f = 2.3$ $m^2/\mu mol$, $a = 1.3$ $\mu mol/m^2$, $b = 0.3$, $c_1 = 0.4$, $c_2 = 0.01$ (1/mM), $c_3 = 0.2$ (1/mM); $k2$ same as $k1$ except $f = 1.3$ (higher ionic strength); $k3$ same as $k1$ except $c_1 = 0.4$ (lower RP analyte retention); $k4$ same as $k1$ except $b = 0.5$ (higher IPR adsorbophilicity).

3.2   EMPIRICAL MODELS OF IPC

The main difference between theoretical and empirical models is that the former are predictive in their own right and theoretical modeling may also be used for finding optimal experimental conditions. Empirical models are characterized by optimization approaches that are mainly interpretative since they are data driven.

3.2.1   CHEMOMETRICAL OPTIMIZATION METHODS

Chemometrical optimization methods were developed because experimental parameters (factors) are interlocking and the required serial one-at-a-time tweaking usually produces too much raw data, from which it is almost impossible to gain a clear knowledge of a system. The chemometrical experimental design represents the planning of experiments aimed at finding the most efficient and clear way to model data based on a limited number of experiments. A treatment represents a set of factors (experimental conditions) decided by the researchers. The experimental objective is usually to obtain a response surface to estimate the influence of the factor levels and their interactions on chromatographic retention (but also resolution, run time, and other parameters). A full two-factor second-order polynomial equation is the most versatile empirical model over a limited domain of factors; the system response (y) may be represented by a surface and can be written:

$$y_i = a_0 + a_1 x_{1i} + a_2 x_{2i} + a_{12} x_{1i} x_{2i} + a_{11} x_{1i}^2 + a_{22} x_{2i}^2 + e_i \qquad (3.32)$$

where x_1 and x_2 are the selected factors, a_0 is an offset parameter, a_1 and a_2 are first-order parameters, a_{12} is an interaction parameter, a_{11} and a_{22} are second-order parameters, and e_1 is a residual, that is, the deviation between the measured system response and that predicted by the model; the subscript i indicates the system response measured by i-th treatments administered to the system by the chromatographer.

Obviously (1) higher order terms may be further included in the equation; (2) more than two factors can be used to describe the system and higher order interaction terms may be present; (3) some terms of Equation 3.32 can, under simpler circumstances, be omitted, for example, in a simple two-factor first-order model, the fourth, fifth, and sixth terms in the right member of Equation 3.32 are missing; and (4) the number of parameters in the model, if k factors are considered, is $(k + 1)(k + 2)/2$; this number must always be lower than the number of treatments; otherwise the fitting parameter optimization is not possible because there are fewer equations than the unknowns.

The goal of the chemometrical experimental design is to explore the response surface and optimize the fitting parameters in Equation 3.32. Their estimates are considered optimized if the residuals are as low as possible. The choice of an experimental design depends on the objectives of the experiment and the number of controlled variables to be investigated; typical IPC factors are IPR concentration, organic modifier percentage, ionic strength, pH, and temperature. Figures 3.5 and 3.6 detail common experimental designs.

The factorial design is one common experimental design; all input factors are set at two (or more) levels each. Curvature in a single factor can be detected only if more than two levels of the factor are administered during system treatments. A design with all possible combinations of all input factors is called a full factorial design in two levels. The use of full factorial designs for optimum parameter selection requires a large number of chromatographic runs since a l-level k-factor full factorial design

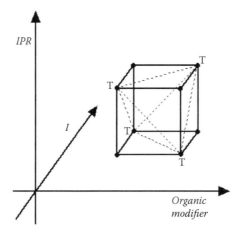

FIGURE 3.5 Two-level–three-factor experimental design with geometrically balanced tetrahedric subset (T).

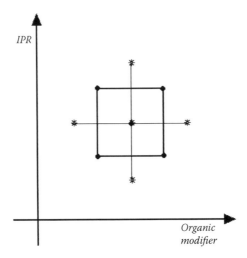

FIGURE 3.6   Central composite design.

requires l^k runs. When only a judiciously (geometrically balanced) chosen fraction of the runs is selected, the result is a fractional factorial design (see Figure 3.5). For example, in a two-level three-factor factorial design with a tetrahedrical subset, only four of eight treatments must be run; the four experimental conditions can only be used to optimize a first-order three-factor model and curvature and interactions cannot be detected:

$$y_i = a_0 + a_1 x_{1i} + a_2 x_{2i} + a_3 x_{3i} + e_i \tag{3.33}$$

The extremely versatile full second-order polynomial model in Equation 3.32 can also be fitted when at least three-level factorial designs are used and 3^k experiments are run. Alternatively, a central composite design may be used effectively (see Figure 3.6). This model is composite because it consists of the overlapping of a star design with $2k + 1$ factor combinations, and a two level k-factor design with 2^k factor combinations to give a total of $2^k + 2k + 1$ treatments.

The distance from the center of the design space to a factorial point for each factor is lower than the distance from the center of the design space to a star point. The central design is so named because the central point is shared by the star design and the factor design and is usually the chromatographer's best guess of optimum conditions. It is usually replicated to estimate pure experimental uncertainty. If curvature and interactions are to be detected, this model is more efficient than a three-level full factorial design when factors are three or more since fewer experiments are required to optimize fitting parameters.

We now illustrate some recent examples of chemometrical modeling of IPC systems for the sake of clarity. In the framework of a quality by design approach, statistically designed experiments were used to optimize the IPC condition for the analysis of atomoxetine and impurities and demonstrate method robustness.

A five-factor (buffer concentration, pH, IPR and organic solvent concentrations, and column temperature) two-level fractional factorial design with four center points was performed. The center points of the design were the midpoints of the range for each factor. The response from the design did not focus on bare retention but on resolution, and particularly on the separation of potential impurities around the main peak. Peak tailing, run time and backpressure were also considered chromatographic responses [76].

The central composite design was often selected because of the limited number of experiments needed to sample the response surfaces. In the separation of As and Se species in tap water, the analysis of isoresponse curves allowed the determination of optimum chromatographic conditions and the robustness of the method [77]. The same design was also used to study the influence of an organic modifier and IPR concentration on retention of biogenic amines in wines. To obtain a compromise between resolution and chromatographic time, optimization through a multi-criteria approach was followed [78].

A central composite design with four factors was devised to optimize the elution gradient method for maize proteins. The resolutions between adjacent peaks and analysis time were selected as six responses. Time and concentration of IPR were important factors. A second-order regression model was employed to calculate the values of the variables that optimized each response [79]. The central composite design proved indispensable for detecting relevant quadratic effects of organic modifier and IPR content in the eluent on retention of heterocyclic aromatic amines [80] and the effects of small changes of chromatographic parameters in the IPC of gentamicin [81]. Several software packages that include eluent composition optimization facilities for liquid chromatography are currently available.

Among chemometrical approaches, the Simplex algorithm [82] involving an evolutionary movement in the factor space to improve the response of interest was applied to analyze antibiotics in pharmaceutical formulations. The selected control variables included the concentrations of the organic modifier and the IPR in the eluent. The response variables were the peak area, resolution, asymmetry factor, and total number of chromatographic peaks. The mobile phase composition that gave optimum results was adopted [83]. A Simplex algorithm was successfully used to maximize the resolution of 14 cosmetic preservatives [84].

If the starting values of the factors are not judiciously chosen, the chemometrical optimization often yields only local optima; moreover its interpretative nature makes predictions out of the studied parameter space unreliable; since it is not theory based, the estimates of the fitting parameters have no physical meaning.

3.2.2 Artificial Neural Network (ANN) Models

Artificial intelligence is related to programs that can learn by introducing new knowledge to a computer and developing a representation of knowledge (expert system) that can be used to attain certain goals. Learning in biological systems involves adjustments to the synaptic connections between the neurons. This is true of ANNs as well. ANN is an artificial intelligence technique (along with genetic algorithms and induction) considered a "soft" retention modeling system because it does not make use of retention equations and theoretical formulae [85].

ANNs have found growing popularity among chromatographers as retention mapping methods; their aim is to describe the chromatographic response surface of analyte retention as a function of several input variables (components of eluent) in the design space. An ANN mimics the current understanding of the brain information processing system postulated to consist of millions of interconnected neurons. ANNs are used to build data-driven models since ANNs like people learn by example during training.

The analyte retention factor is measured at a number of mobile phase compositions (design points); the response surface is modelled and then the retention factor can be predicted at every eluent composition in the design space. ANNs are not programmed to solve specific problems because that would require the programmer to know the problem solution. ANNs can manage problems that chromatographers cannot solve. An ANN consists of a great number of single processing elements called neurons or nodes, organized in layers. They are linked to each other and work in unison to solve specific problems. An artificial neuron is a device involving several inputs and a single output. Usually each input (x_i) is weighted, that is, it is multiplied with a number (w_i) which, along with the bias of the neuron ($bias_i$), yields the weighted input. If the sum of these weighted inputs exceeds a pre-set threshold value T, the artificial neuron, like a human one, fires and produces an output (y_i) that can vary continuously as the inputs (linear or sigmoid neurons) change or can be set only at a fixed level (threshold neuron).

$$y_i = f\left(\sum_i x_i w_i + bias_i\right) \qquad (3.34)$$

During the training mode, a neuron can be educated to fire (or not) for particular input patterns. In the using mode, when a taught input pattern is detected, its associated output becomes the current output. If the input pattern does not belong in the taught list of input patterns, the firing rule, based on sense of similarity with a taught input pattern, is used to determine whether to fire or not; this gives the neuron a kind of sensibility to pattern not seen during training. Its learning is adaptive because weights and/or thresholds can be changed via several algorithms. At variance with feed-forward networks, the feedback systems are more powerful and complicated because signals travel in both directions via loops introduced in the network. They are dynamic since their condition evolves continuously until they attain an equilibrium point.

A new equilibrium must be found only when the input changes. The architecture of an ANN is usually multi-layer. At minimum, a hidden layer is placed between the input and output layers. Every neural network possesses knowledge contained in the values of the connection weights. Modifying the knowledge stored in the network as a function of experience implies a learning rule for changing the values of the weights.

The prediction power of theoretical modeling and an ANN model in IPC were compared. A feed-forward layered back-propagation ANN was used; the input layer consisted of three neurons representing the fraction of organic modifier, the IPR, and counter ion concentrations. The number of neurons in the hidden layer was optimized by lateral inhibition. The number of output neurons equaled the number of analytes in the sample. The output neuron for the i-th analytes yielded a predicted retention factor that was a function of the input pattern and the weights. The system computed the error derivative of the weights. Learning is the changing of the weights during the training mode, until the difference between the predicted outputs and the measured retention factors is minimized. The study demonstrated that theoretical modeling is the most eligible choice when few experimental data are available, while the neural network software accounted for a greater versatility. The power of both models was similar for large numbers of data [86].

In the previous paragraph we explained that experimental design is useful to model the response surface described by an algebraic expression and also proved valuable when ANNs were used to model chromatographic responses. An optimization by experimental design and ANN of the IPC separation of 20 cosmetic preservatives correlated the retention time of each analyte to the system variables and their interactions. The ANN showed good predictive ability and was used by a grid search algorithm to optimize chromatographic conditions for the separation [87]. A fractional factorial design and a particular central composite (Hoke) design were used along with ANN modeling to optimize the resolution and total analysis time during IPC testing of five pesticides [88]. It is worth noting that ANN models of IPC separations are limited but may become widespread in the years ahead.

REFERENCES

1. Knox, J.H. and Hartwick, R.A. Mechanism of ion-pair liquid chromatography of amines, neutrals, zwitterions and acids using anionic hetaerons. *J. Chromatogr.* 1981, 204, 3–21.
2. Gloor, R. and Johnson, E.L. Practical aspects of reversed phase ion-pair chromatography. *J. Chromatogr. Sci.* 1977, 15, 413–423.
3. Horvath, C. et al. Enhancement of retention by ion-pair formation in liquid chromatography with nonpolar stationary phases. *Anal. Chem.* 1977, 49, 2295–2305.
4. Melander, W.R. and Horvath, C. Mechanistic stufy of ion-pair reversed-phase chromatography. *J. Chromatogr.* 1980, 201, 211–224.
5. Wittmer, D.P., Nuessle, N.O., and Haney, W.G., Jr. Simultaneous analysis of tartrazine and its intermediates by reversed phase liquid chromatography. *Anal. Chem.* 1975, 47, 1422–1423.
6. Knox, J.H. and Laird, G.R. Soap chromatography: a new high performance chromatographic technique for separation of ionizable materials: dyestuff intermediates. *J. Chromatogr.* 1976, 122, 17–34.
7. Kraak, J.C., Jonker, K.M., and Huber, J.F.K. Solvent-generated ion exchange systems with anionic surfactant for rapid separation of amino acids. *J. Chromatogr.* 1977, 142, 671–688.
8. Ghaemi Y. and Wall R.A. Hydrophobic chromatography with dynamically coated stationary phases. *J. Chromatogr.* 1979, 174, 51–59.
9. Fornstedt, T., Peak distortion effects of suramin due to large system peaks in bioanalysis using ion-pair adsorption chromatography. *J. Chromatogr. B.* 1993, 612, 137–144.
10. Hoffman, N.E. and Liao, J.C. Reversed phase high performance liquid chromatographic separations of nucleotides in the presence of solvophobic ions. *Anal. Chem,* 1977, 49, 2231–2234.

11. Kissinger, P.T. Comments on reversed-phase ion-pair partition chromatography. *Anal. Chem.* 1977, 49, 883–883.
12. Iskandarani, Z. and Pietrzyk, D. J. Ion interaction chromatography of organic anions in the presence of tetraalkylammonium salts. *Anal. Chem.* 1982, 54, 1065–1071.
13. Xianren, Q. and Baeyens, W. Retention and separation of inorganic anions by reversed-phase ion-interaction chromatography on octadecyl silica. *J. Chromatogr.* 1988, 456, 267–285.
14. Hung, C.T. and Taylor, R.B. Mechanism of retention of acidic solutes by octadecyl silica using quaternary ammonium pairing ions as ion exchangers. *J. Chromatogr.* 1980, 202, 333–345.
15. Bidlingmeyer, B.A. et al. Retention mechanism for reversed-phase ion-pair liquid chromatography. *J. Chromatogr.* 1979, 186, 419–434.
16. Bidlingmeyer, B.A. Separation of ionic compounds by reversed-phase liquid chromatography: an update of ion-pairing techniques. *J. Chromatogr.* 1980, 18, 525–539.
17. Sarzanini, C. et al., Retention model for anionic, neutral and cationic analytes in reversed-phase ion interaction chromatography. *Anal. Chem.* 1996, 68, 4494–4500.
18. LoBrutto, R., Jones, A., and Kazakevich, Y.V. Effect of counter-anion concentration on retention in high-performance liquid chromatography of protonated basic analytes. *J. Chromatogr. A.* 2001, 913, 189–196.
19. Bartha, A. and Ståhlberg , J. Electrostatic retention model of reversed-phase ion-pair chromatography. *J. Chromatogr. A.* 1994, 668, 255–284.
20. Cecchi, T., Pucciarelli, F., and Passamonti, P. Extended thermodynamic approach to ion-interaction chromatography. *Anal. Chem.* 2001, 73, 2632–2639.
21. Kong, R.C., Sachok, B., and Deming, S.N. Combined effects of pH and surface-active-ion concentration in reversed-phase liquid chromatography. *J. Chromatogr.* 1980, 199, 307–316.
22. Stranahan, J.J. and Deming, S.N. Thermodynamic model for reversed-phase ion-pair liquid chromatography. *Anal. Chem.* 1982, 54, 2251–2256.
23. Zou, J., Motomi, S., and Fukutomi, H. Reversed-phase ion-interaction chromatography of inorganic anions with tetraalkylammonium ions and divalent organic anions using indirect photometric detection. *Analyst* 1991, 116, 1399–1405.
24. Ståhlberg, J. and Bartha, A. Extension of the electrostatic theory of reversed-phase ion-pair chromatography for high surface concentrations of the adsorbing amphiphilic ion. *J. Chromatogr. A.* 1988, 456, 253–265.
25. Ståhlberg, J. The Gouy-Chapman theory in combination with a modified Langmuir isotherm as a theoretical model for ion-pair chromatography. *J. Chromatogr.* 1986, 356, 231–245.
26. Cecchi T. Extended thermodynamic approach to ion-interaction chromatography: a thorough comparison with the electrostatic approach and further quantitative validation. *J. Chromatogr. A.* 2002, 958, 51–58.
27. Cecchi T. Use of lipophilic ion adsorption isotherms to determine the surface area and the monolayer capacity of a chromatographic packing, as well as the thermodynamic equilibrium constant for its adsorption. *J. Chromatogr. A.* 2005, 1072, 201–206.
28. Narkiewicz-Michalek, J. Electrostatic theory of ion-pair chromatography: multi-site occupancy model for modifier and solute adsorption. *Chromatographia* 1993, 35, 527–538.
29. Cantwell, F.F. Retention model for ion-pair chromatography based on double-layer ionic adsorption and exchange. *Pharm. Biomed. Anal.* 1984, 2, 153–164.
30. Stern, O. Zur Theorie der Elektrolytischen Doppelschicht. *Z. Electrochem.* 1924, 30, 508–516.
31. Chen, J.G. et al. Electrical double layer models of ion-modified (ion -pair) reversed-phase liquid chromatography. *J. Chromatogr. A.* 1993, 656, 549–576.

32. Liu, H. and Cantwell, F.F. Electrical double-layer model for sorption of ions on octadecylsilyl bonded phases including the role of residual silanol groups. *Anal. Chem.* 1991, 63, 993–1000.

33. Liu, H. and Cantwell, F.F. Electrical double-layer model for ion-pair chromatographic retention on octadecylsilyl bonded phases. *Anal. Chem.* 1991, 63, 2032–2037.

34. Dai, J. and Carr, P.W. Role of ion-pairing in anionic additive effects on the separation of cationic drugs in reversed-phase liquid chromatography. *J. Chromatogr. A.* 2005, 1072, 169–184.

35. Mbuna, J. et al. Capillary zone electrophoretic studies of ion association between nonorganic anions and tetraalkylammonium ions in aqueous dioxane media. *J. Chromatogr. A.* 2005, 1069, 261–270.

36. Dai, J. et al. Effect of anionic additive type on ion-pair formation constants of basic pharmaceuticals. *J. Chromatogr. A.* 2005, 1069, 225–234.

37. Steiner, S.A., Watson, D.M., and Fritz, J.S. Ion association with alkylammonium cations for separation of anions by capillary electrophoresis. *J. Chromatogr. A.* 2005, 1085, 170–175.

38. Motomizu, S. and Takayanagi, T. Electrophoretic mobility study on ion–ion interactions in an aqueous solution. *J. Chromatogr. A.* 1999, 853, 63–69.

39. Takayanagi, T., Wada, E., and Motomizu, S. Electrophoretic mobility study of ion association between aromatic anions and quaternary ammonium ions in aqueous solution. *Analyst* 1997, 122, 57–62.

40. Takayanagi, T., Wada, E., and Motomizu, S. Separation of divalent aromatic anions by capillary zone electrophoresis using multipoint ion association with divalent quaternary ammonium ions. *Analyst* 1997, 122, 1387–1392.

41. Takayanagi, T., Tanaka, H., and Motomizu, S. Ion association reaction between divalent anionic azo dyes and hydrophobic quaternary ammonium ions in aqueous solution as studied by capillary zone electrophoresis. *Anal. Chem.* 1997, 13, 11–18.

42. Popa, T.V., Mant, C.T., and Hodges, R.S. Capillary electrophoresis of amphipathic α-helical peptide diastereomers. *Electrophoresis* 2004, 25, 94–107.

43. Popa, T.V., Mant, C.T., and Hodges, R.S. Ion interaction capillary zone electrophoresis of cationic proteomic peptide standards. *J. Chromatogr. A.* 2006, 1111, 192–199.

44. Popa, T.V. et al. Capillary zone electrophoresis of α-helical diastereomeric peptide pairs with anionic ion-pairing reagents. *J. Chromatogr. A.* 2004, 1043, 113–122.

45. Fini, A. et al. Formation of ion-pairs in aqueous solutions of diclofenac salts. *Int. J. Pharm.* 1999, 187, 163–173.

46. Takeda, Y. et al. Extraction of sodium and potassium picrates with 16-crown-5 into various diluents: elucidation of fundamental equilibria determining the extraction selectivity for Na^+ over K^+. *Talanta* 1999, 48, 559–569.

47. Takeda, Y., Ikeo, N., and Sakata, N. Thermodynamic study of solvent extraction of 15-crown-5- and 16-crown-6-s-block metal ion complexes and tetraalkylammonium ions with picrate anions into chloroform. *Talanta* 1991, 38, 1325–1333.

48. Mbuna, J. et al. Evaluation of weak ion association between tetraalkylammonium ions and inorganic anions in aqueous solutions by capillary zone electrophoresis. *J. Chromatogr. A.* 2004, 1022, 191–200.

49. Lu, P., Zou, H., and Zhang, Y. The retention equation in reversed phase ion-pair chromatography. *Mikrochim. Acta* 1990, III, 35–53.

50. Cecchi, T., Pucciarelli, F., and Passamonti, P. Ion interaction chromatography of neutral molecules. *Chromatographia* 2000, 53, 27–34.

51. Cecchi, T., Pucciarelli, F., and Passamonti, P. An extended thermodynamic approach to ion-interaction chromatography for high surface potential: use of a potential approximation to obtain a simplified retention equation. *Chromatographia* 2001, 54, 589–593.

52. Cecchi, T., Pucciarelli, F., and Passamonti, P. Extended thermodynamic approach to ion-interaction chromatography for low surface potential: use of a linearized potential expression. *J. Liq. Chromatogr. Rel. Technol.* 2001, 24, 2551–2557.

53. Cecchi, T., Pucciarelli, F., and Passamonti, P. Ion Interaction chromatography of neutral molecules: a potential approximation to obtain a simplified retention equation. *J. Liq. Chromatogr. Rel. Technol.* 2001, 24, 291–302.

54. Cecchi, T. et al. *Chromatographia* The dipole approach to ion interaction chromatography of zwitterions. *Chromatographia* 2001, 54, 38–44.

55. Cecchi, T. and Cecchi, P. The dipole approach to ion interaction chromatography of zwitterions: use of a potential approximation to obtain a simplified retention equation. *Chromatographia* 2002, 55, 279–282.

56. Cecchi, T. and Cecchi, P. The dipole approach to ion interaction chromatography of zwitterions: use of the linearized potential expression for low surface potential. *J. Liq. Chromatogr. Rel. Technol.* 2002, 25, 415–420.

57. Cecchi, T., Pucciarelli, F., and Passamonti, P. Extended thermodynamic approach to ion-interaction chromatography: influence of organic modifier concentration. *Chromatographia* 2003, 58, 411–419.

58. Cecchi, T., Pucciarelli, F., and Passamonti, P. Extended thermodynamic approach to ion-interaction chromatograph: effect of the electrical charge of the solute ion. *J. Liq. Chromatogr. Rel. Technol.* 2004, 27, 1–15.

59. Cecchi, T., Pucciarelli, F., and Passamonti, P. Extended thermodynamic approach to ion interaction chromatography: a mono- and bivariate strategy to model the influence of ionic strength. *J. Sep. Sci.* 2004, 27, 1323–1332.

60. Cecchi, T. Influence of the chain length of the solute ion: a chromatographic method for the determination of ion-pairing constants. *J. Sep. Sci.* 2005, 28, 549–554.

61. Cecchi, T., Pucciarelli, F., and Passamonti, P. Ion interaction chromatography of zwitterions: fractional charge approach to model the influence of the mobile phase concentration of the ion-interaction reagent. *Analyst* 2004, 129, 1037–1042.

62. Cecchi, T. et al. The fractional charge approach in ion-interaction chromatography of zwitterions: influence of the stationary phase concentration of the ion interaction reagent and pH. *J. Liq. Chromatogr. Rel. Technol.* 2005, 28, 2655–2667.

63. Cecchi, T. Comprehensive thermodynamic approach to ion interaction chromatography. In *Dekker Encyclopedia of Chromatography*. Cazes, J., Ed., Marcel Dekker: New York, 2004.

64. Cecchi, T. and Passamonti, P. Retention mechanism for ion-pairing chromatography with chaotropic additives. *J. Chromatogr. A.* 2009, 1216, 1789–1797, and Erratum in *J. chromatogr. A.* 2009, 1216, 5164.

65. Davies, J.T. and Rideal, E.K. *Adsorption at Liquid Interfaces: Interfacial Phenomena.* Academic Press: New York, 1961, pp. 154–216.

66. Rosen, M.J. Adsorption of surface-active agents at interfaces: the electrical double layer. In *Surfactant and Interfacial Phenomena*, 2nd ed., Wiley: New York, 1978, pp. 33–106.

67. Crommen, J., Fransson, B., and Shill, G. Ion-pair chromatography in the low concentration range by use of highly absorbing counter-ions. *J. Chromatogr.* 1977, 142, 283–297.

68. Yoshida, Y. et al. Evaluation of distribution ratio in ion-pair extraction using fundamental thermodynamic quantities. *Anal. Chim. Acta* 1998, 373, 213–225.

69. Yoshikawa, Y., Terada, H. Ion-pair partition mechanism of methyl orange with aminoalkanols and alkylamines. *Chem. Pharm. Bull.* 1994, 42, 2407–2411.

70. Shoenmakers, P.J., Billiet, A.H., and De Galan L. Influence of organic modifiers on the retention behaviour in reversed-phase liquid chromatography and its consequences for gradient elution. *J. Chromatogr.* 1979, 185, 179–195.

71. Zappoli, S. and Bottura, C. Interpretation of the retention mechanism of transition metal cations in ion interaction chromatography. *Anal. Chem.* 1994, 66, 3492–3499.
72. Melander, W.R., Corradini, D., and Horváth, C. Salt-mediated retention of proteins in hydrophobic-interaction chromatography: application of solvophobic theory. *J. Chromatogr.* 1984, 317, 67–85.
73. Jandera, P., Churacek, J., and Taraba, B. Comparison of retention behaviour of aromatic sulphonic acids in reversed-phase systems with mobile phases containing ion-pairing ions and in systems with solutions of inorganic salts as the mobile phases. *J. Chromatogr.* 1983, 262, 121–140.
74. Rafferty, J.L. et al., Retention mechanism in reversed-phase liquid chromatography: molecular perspective. *Anal. Chem.* 2007, 79, 6551–6558.
75. Prigogine, I. *Introduction to Thermodynamics of Irreversible Processes*, 3rd ed., Wiley: New York, 1967.
76. Gavin, P.F. and Olsen, B.A. A quality by design approach to impurity method development for atomoxetine hydrochloride (LY139603). *J. Pharm. Biomed.* 2008, 46, 431–441.
77. Do, B. et al. Application of central composite designs for optimisation of the chromatographic separation of monomethylarsonate and dimethylarsinate and of selenomethionine and selenite by ion-pair chromatography coupled with plasma mass spectrometric detection. *Analyst* 2001, 126, 594–601.
78. Hlabangana, L., Hernandez-Cassou, S., and Saurina J. Determination of biogenic amines in wines by ion-pair liquid chromatography and post-column derivatization with 1,2-naphthoquinone-4-sulphonate. *J. Chromatogr. A.* 2006, 1130, 130–136.
79. Rodriguez-Nogales, J.M., Garcia, M.C., and Marina, M.L. Development of a perfusion reversed-phase high performance liquid chromatography method for the characterisation of maize products using multivariate analysis. *J. Chromatogr. A.* 2006, 1104, 91–99.
80. Bianchi, F. et al. Investigation of the separation of heterocyclic aromatic amines by reversed phase ion-pair liquid chromatography coupled with tandem mass spectrometry: the role of ion-pair reagents on LC-MS/MS sensitivity. *J. Chromatogr. B.* 2005, 825, 193–200.
81. Manyanga, V. et al. Improved liquid chromatographic method with pulsed electrochemical detection for the analysis of gentamicin, *J. Chromatogr. A.* 2008, 1189, 347–354.
82. Cela, R. et al. PREOPT-W: off-line optimization of binary gradient separation in HPLC by simulation IV: phase 3. *Comput. Chem.* 1996, 20, 315–330.
83. Sarri, A.K., Megoulas, N.C., and Koupparis M.A. Development of novel liquid chromatography evaporative light scattering detection method for bacitracins and applications to quality control of pharmaceuticals. *Anal. Chim. Acta* 2006, 573–574, 250–257.
84. Marengo, E., Gennaro M.C., and Gianotti, V. A Simplex-optimized chromatographic separation of 14 cosmetic preservatives: analysis of commercial products. *J. Chromatogr. Sci.* 2001, 39, 399–344.
85. Haykin, S. *Neural Networks: A Comprehensive Foundation*, 2nd ed. Prentice Hall: New Jersey, 1999.
86. Sacchero, G. et al. Comparison of prediction power between theoretical and neural-network models in ion-interaction chromatography. *J. Chromatogr. A.* 1998, 799, 35–45.
87. Marengo, E. et al. Optimization by experimental design and artificial neural networks of the ion-interaction reversed-phase liquid chromatographic separation of 20 cosmetic preservatives. *J. Chromatogr. A.* 2004, 1029, 57–65.
88. Marengo, E., Gennaro, M.C., and Angelino, S. Neural networks and experimental design to investigate the effects of five factors in ion-interaction high-performance liquid chromatography. *J. Chromatogr. A.* 1998, 799 47–55.

4 Modeling of Retention as a Function of Analyte Nature

Modeling of analyte retention as a function of the mobile phase composition has been the focus of interest of chromatographers since the infancy of ion-pairing chromatography (IPC). Despite the widespread use of the technique, the correlation and prediction of the retention factor (k) as a function of analyte nature at constant mobile phase composition has only recently been initiated [1–3]. This advance makes use of quantitative structure retention relationships (QSRRs) to identify the properties of analytes that control their retention. The retention factors of a set of analytes are measured at certain experimental conditions and multiple regression analysis is used to correlate k with several solute molecular descriptors such as physicochemical, topological, geometrical, and electronic factors. One kind of QSRR is based on the so-called linear solvation energy relationships (LSERs).

The first effort to use LSERs in IPC relied on a retention equation based on a mixture of stoichiometric and electrostatic models. Several approximations were made [1–3]. First, ion-pairing in the eluent was neglected, but this is at variance with clear qualitative and quantitative experimental results [4–13]. In Chapter 3 (Section 3.1.1), the detrimental consequences of this assumption were clarified and demonstrated that extensive experimental evidence cannot be rationalized if pairing interactions in the eluent are not taken into account. Furthermore, in the modeling of k as a function of the analyte nature, the presence of the IPR in the eluent was assumed not to influence the retention of neutral analytes. This assumption is only occasionally true [14,15] and the extended thermodynamic retention model of IPC suggests the quantitative relationship between neutral analytes retention and IPC concentration in the eluent [16].

Moreover, the only solute descriptor for ion-pair effect was the analyte charge, but it was shown that the analyte charge status did not explain (1) different experimental curves when k is plotted as a function of the stationary phase concentration of the IPR for various IPRs; (2) the dependence of the ratio of the retention of two different analytes on IPR concentration; (3) the dependence of the k/k_0 ratio on the analyte nature if experimental conditions are the same [16,17]; and (4) ion-pairing of peptides [12]. The model makers realized that the charge may have been a too-simple solute descriptor for ion-pairing because it did not exhibit the hydrophobic effect, but they did not devise a better descriptor. Section 2.5.3 (Chapter 2) details the peculiarities of the ion-pairing process of hydrophobic ions and it is clear that interactions other than electrostatic are essential to quantitatively describe the duplex formation that is not even conceivable in aqueous medium according to the electrostatic model. However, behaviors at variance with electrostatic predictions were common. Consequently, further solute descriptors should be introduced.

Finally, the fitting coefficients of the solvation parameter model, determined using neutral solutes in the initial linear regression, were considered fixed coefficients in the subsequent nonlinear least-squares fitting to achieve consistency upon the inclusion of ionized solutes. Hence the trustworthiness and reliability of the coefficients was actually compelled and not confirmed. The major concern that arises upon analyzing the attempts of Li and co-workers to use LSER in IPC is the absence of an interpretation of the ion-pair parameter estimates. This fitting variable has a clear physical meaning and an estimate is crucial for judging the adequacy of the model.

The authors presumed that the estimate must have been zero when phosphoric acid was used as the IPR but they did not test their guess. A theoretical analysis of the meaning of this parameter (Equation 9 in Reference 1) confirms that the value cannot be zero. A phosphate ion was used as an IPR [18] and was demonstrated to be effective [11,19–23]. Unfortunately, the estimates of this variable obtained for different IPRs are not in agreement with the theoretical values. If one keeps in mind that enough processing of large quantities of data with a sufficient number of fitting parameters may lead to any deduction [24], it can be concluded that the use of QSRRs in IPC is still in its infancy.

The influences of individual characteristics of several sulfonated azo dyes on IPC retention were considered and much easier solute descriptors, i.e., lipophilic and polar indices, were adopted [25]. While the polar index linearly increases with increasing numbers of sulfonic acid groups in the azo dye molecule, the lipophilic index is linearly dependent on the molecular weight of the dye. These descriptors, interestingly, depended slightly on the number of carbon atoms in the IPR, thereby confirming that the pairing mechanism is not only charge driven. Structural effects can be used to make an educated guess about the architecture of an unknown analyte on the basis of chromatographic data or predict retention of known analytes under specific experimental conditions.

The separation selectivities of C18 columns from different manufacturers were compared to evaluate the applicability of sequence-specific retention calculator peptide retention prediction algorithms. Pore size was found to be the main factor affecting selectivity, while differences in end-capping chemistry did not play a major role. The introduction of embedded polar groups to the C18 functionality (see Chapter 5 for more details on this topic) enhanced the retention of peptides containing hydrophobic amino acid residues with polar groups. It was also demonstrated that changing the IPR significantly reduced algorithm predictive ability [26].

In a study of retention of aromatic carboxylic acids under IPC conditions, linear free energy relationships were observed between the capacity factors and the extraction equilibrium constants of benzoic acid and naphthalene carboxylic acid. The capacity factor of benzene polycarboxylic acids was directly related to their association constants and quaternary ammonium ions calculated on the basis of an electrostatic interaction model [27,28].

Future theoretical efforts to model retention as a function of solute structure are needed. Obviously the thermodynamically sound description of the pairing process is a prerequisite to meeting this goal. Competent chromatographers very much doubted that the ability to quantitatively predict experimental conditions to achieve an acceptable separation is an achievable goal because changes in the free energy of adsorption as small as a few hundredths of RT are chromatographically important [29].

REFERENCES

1. Li, J. Prediction of internal standards in reversed-phase liquid chromatography IV: correlation and prediction of retention in reversed-phase ion-pair chromatography based on linear solvation energy relationships. *Anal. Chim. Acta* 2004, 522, 113–126.
2. Li, J. and Rethwill, P.A. Prediction of internal standards in reversed phase liquid chromatography. *Chromatographia* 2004, 60, 63–71.
3. Li, J. and Rethwill, P.A. Systematic selection of internal standard with similar chemical and UV properties to drug to be quantified in serum samples. *Chromatographia* 2004, 60, 391–397.
4. Dai, J. et al. Effect of anionic additive type on ion-pair formation constants of basic pharmaceuticals. *J. Chromatogr. A.* 2005, 1069, 225–234.
5. Steiner, S.A., Watson, D.M. and Fritz, J.S. Ion association with alkylammonium cations for separation of anions by capillary electrophoresis. *J. Chromatogr. A.* 2005, 1085, 170–175.
6. Motomizu, S. and Takayanagi, T. Electrophoretic mobility study on ion–ion interactions in an aqueous solution. *J. Chromatogr. A.* 1999, 853, 63–69.
7. Takayanagi, T., Wada, E., and Motomizu, S. Electrophoretic mobility study of ion association between aromatic anions and quaternary ammonium ions in aqueous solution. *Analyst* 1997, 122, 57–62.
8. Takayanagi, T., Wada, E., and Motomizu, S. Separation of divalent aromatic anions by capillary zone electrophoresis using multipoint ion association with divalent quaternary ammonium ions. *Analyst* 1997, 122, 1387–1392.
9. Takayanagi, T., Tanaka, H., and Motomizu, S. Ion Association Reaction between Divalent Anionic Azo Dyes and Hydrophobic Quaternary Ammonium Ions in Aqueous Solution as Studied by Capillary Zone Electrophoresis. *Anal. Sci.* 1997, 13, 11.
10. Popa, T.V., Mant, C.T., and Hodges, R.S. Capillary electrophoresis of amphipathic α-helical peptide diastereomers. *Electrophoresis* 2004, 25, 94–107.
11. Popa, T.V., Mant, C.T., and Hodges, R.S. Ion-interaction capillary zone electrophoresis of cationic proteomic peptide standards. *J. Chromatogr. A.* 2006, 1111, 192–199.
12. Popa, T.V. et al. Capillary zone electrophoresis of α-helical diastereomeric peptide pairs with anionic ion-pairing reagents. *J. Chromatogr. A.* 2004, 1043, 113–122.
13. Mbuna, J., Takayanagi, T., Oshima, M., Motomizu, S. *J. Chromatogr. A.* Evaluation of weak ion association between tetraalkylammonium ions and inorganic amions in aqueous solutions by capillary zone electrophoresis. 2004, 1022, 191–200.
14. Cecchi, T., Pucciarelli, F., and Passamonti, P. Ion interaction chromatography of neutral molecules. *Chromatographia* 2000, 53, 27–34.
15. Cecchi, T., Pucciarelli, F., and Passamonti, P. Ion interaction chromatography of neutral molecules: a potential approximation to obtain a simplified retention equation. *J. Liq. Chromatogr. Rel. Technol.* 2001, 24, 291–302.
16. Cecchi, T., Pucciarelli, F., and Passamonti, P. Extended thermodynamic approach to ion interaction chromatography. *Anal. Chem.* 2001, 73, 2632–2639.
17. Cecchi T. Comprehensive thermodynamic approach to ion interaction chromatography. In *Dekker Encyclopedia of Chromatography,* Cazes, J., Ed. Marcel Dekker: New York, 2004.

18. Kallel, L. et al. Optimization of the separation conditions of tetracyclines on a preselected reversed-phase column with embedded urea group. *J. Sep. Sci.* 2006, 29, 929–935.

19. Hashem, H. and Jira, T. Effect of chaotropic mobile phase additives on retention behaviour of β-blockers on various reversed-phase high-performance liquid chromatography columns. *J. Chromatogr. A.* 2006, 1133, 69–75.

20. Basci, N.E. et al. Optimization of mobile phase in the separation of β-blockers by HPLC. *J. Pharm. Biomed. Anal.* 1998, 18, 745–750.

21. LoBrutto, R. et al. Effect of the eluent pH and acidic modifiers in high-performance liquid chromatography retention of basic analytes. *J. Chromatogr. A.* 2001, 913, 173–187.

22. Mant, C.T. and Hodges, R.S. Context-dependent effects on the hydrophilicity and hydrophobicity of side chains during reversed-phase high-performance liquid chromatography: implications for prediction of peptide retention behaviour. *J. Chromatogr. A.* 2006, 1125, 211–219.

23. Zhang, L. et al. Simultaneous determination of baicalin, baicalein, wogonin, oxysophocarpine, oxymatrine and matrine in the Chinese herbal preparation of Sanwu-Huangqin-Tang by ion-paired HPLC. *Chromatographia* 2007, 66, 115–120.

24. Lavine, B. and Workman, J. Chemometrics. *Anal. Chem.* 2006, 78, 4137–4145.

25. Vanérková, D., Jandera, P., and Hrabica, J. *J. Chromatogr. A.* Behaviour of sulphonated azo dyes in ion-pairing reversed-phase high-performance liquid chromatography *J. Chromatogr. A.* 2007, 1143, 112–120.

26. Spicer, V. et al. Sequence-specific retention calculator: a family of peptide retention time prediction algorithms in reversed-phase HPLC: applicability to various chromatographic conditions and columns. *Anal. Chem.* 2007, 79, 8762–8768.

27. Kawamura, K. et al. Separation of aromatic carboxylic acids using quaternary ammonium salts on reversed-phase HPLC 2: application for the analysis of Loy Yang coal oxidation products. *J. Sep. Sci. Technol.* 2006, 41, 723–732.

28. Kawamura, K. et al. Separation of aromatic carboxylic acids using quaternary ammonium salts on reversed-phase HPLC 1: separation behavior of aromatic carboxylic acids. *J. Sep. Sci. Technol.* 2006, 41, 79–390.

29. Vitha, M. and Carr, P.W. The chemical interpretation and practice of linear solvation energy relationships in chromatography. *J. Chromatogr. A.* 2006, 1126, 143–194.

5 Stationary Phases

Chromatographic stationary phases are usually obtained via the correct attachment of proper ligands to a suitable support, thereby obtaining a bonded phase. The characteristics of both the support and the ligands exert profound impacts on the performance of an IPC separation. Key parameters are the physical features of the support (particle size, pore diameter, specific surface area), the chemical nature of the bonded phase, and its bonding density.

Most chromatographic stationary phases are silica-based. The shapes of silica surfaces are basically unknown. The only method that allows direct measurement of the outer surface shape is atomic force microscopy. Silica-based adsorbents exhibit outstanding mechanical resistance and can provide exceptional efficiency; ultrapure silicas are now available. The specific surface area usually ranges from 120 to 450 m^2/g and pore diameters from 60 to 150 Å; the specific pore volume (volumes of all pores in 1 g adsorbent) is 0.5 to 1.2 ml/g.

The main downside of silica-based sorbents is pH compatibility: they are stable only at pH values from 2 to 8; at lower pH, the H$_3$O$^+$ ions catalyze the hydrolysis of the alkyl ligand; at higher pH, the nucleophilic attack of Si-O bonds by OH$^-$ results in the erosion of the silica surface. The formation of Si(OH)$_4$ is signaled by a back pressure increase. Recently the chemical resistance of the silica particles has been improved by hybrid particle technology. Organic moieties such as ethyl groups were inserted close to the particle surfaces and columns stable between pH 1 and 12 under ultra-high pressure (1000 bar) were produced.

Polymer-based stationary phases suffer from lower mechanical resistance, even if their pH stability is impressive. The following section focuses on the physical description of silica-based adsorbents since they are the most common.

Stationary phases are distinguished in reversed and normal phases based on the hydrophobic or polar chemical nature of the ligand, respectively. Alkyl chains of assorted lengths (eg. octyl-c8-, octadecyl-C18-) are characteristic of reversed phase materials, while normal phase packings are functionalized with chains bearing –OH, –CN, or –NH$_2$ groups. Reversed phase chromatography exploits non-specific dispersive forces between the analyte and the stationary phase surface: they are weaker than polar interactions, but allow a fine tuning of retention since they depend on the extension and architecture of the molecules. It follows that the reversed phase selectivity is higher and almost 80% of all HPLC separations are performed under reversed phase conditions.

Since the discovery of IPC, this separation strategy has been devoted to increasing the mediocre retention of ionized samples on reversed phase stationary phases via ion-pair formation with a suitable IPR to increase the analyte hydrophobicity and, in turn, its retention. It is therefore clear that most IPC separations are performed under reversed phase conditions. Even if normal phase chromatography exploits polar interactions (dipole–dipole, dipole–induced dipole) between analytes and the polar stationary phase, we will highlight a rare example of this mode of IPC.

5.1 REVERSED PHASE IPC

5.1.1 SILICA-BASED STATIONARY PHASES AND SILANOPHILIC INTERACTIONS

The hydrophobicity of the stationary phase packing is a crucial factor because it influences the retention of both the IPR and the solute ion; both retentions are greater when the ligand alkyl chain length is longer and the ligand bonding density of the bonded phase is higher. The most common bonded phases are of the C18 type.

In Chapter 3 (Section 3.1.2), we detailed a theory that may provide a rationalization of these effects. If the hydrophobicity of the ligand site (L) rises, the adsorption free energy is more negative (spontaneity of the process increases) and $K_{LE,}$, K_{LH}, and K_{LEH} are all higher. Increased adsorption of the IPR involves stronger electrostatic surface potential that improves oppositely charged analyte retention. These theoretical considerations explain why different alkyl chain lengths were employed along with the most common C18 and C8 silica packings. For example, a C30 packing material was used for the IPC of inorganic ions [1] because only the high hydrophobicity of such a bonded phase afforded adequate retention of the hydrophilic analytes. Conversely, a C4 column proved valuable for reducing the chromatographic cycle time including regeneration for gradient IPC of amino acids [2]. A phenylhexyl phase was used for the IPC of heterocyclic aromatic amines [3].

Even if chain length is the key parameter, the ligand bonding density (usually above 2.5 μmol/m^2) may be very influential in determining overall stationary phase hydrophobicity. When the monolayer capacity, theoretically estimated by $[L]_T$, increases as a result of increased bonding density, adsorption competitions are less operative and enhanced retention is expected. It should be noted that ligand bonding density can be calculated on the basis of the column carbon load and the total surface area of the column.

The former represent the weight percent of carbon atoms on the adsorbent sample measured by elemental analysis. The bonding density of a given carbon load decreases with increasing surface area of the silica support; porosity multiplies the available particle surface. The specific surface area (surface area in 1 g adsorbent) is said to be inversely proportional to pore diameter (at constant specific pore volume) and obviously increases with increasing specific pore volume.

The estimate of the surface area of chromatographic silica support is a complicated issue. It is usually performed via the BET method using low temperature nitrogen adsorption (N$_2$.- sorptometry). The total surface area of the adsorbent is the product of the number of adsorbed molecules and the surface area per molecule. However, if the pore size distribution is not very narrow, an estimate of bonding density on the basis of carbon load and surface area may yield a large error because the smallest pores are not available for derivatization and the calculated bonding density is lower than the actual one.

Pore diameter sanctions the ability of analyte molecules to penetrate inside the particle and interact with its inner surface. Small pores (less than 10 ÷ 50 Å) should be absent because the slow mass transfer inside them impairs column efficiency. Furthermore, pore diameters are crucial for the correct theoretical modeling (see Section 3.1.2) of analyte retention as a function of IPR concentration. According to Weber [4], the semi-infinite geometry does not properly apply if the pore size is of the same order of magnitude as the Debye length. If the pore diameter is small (<100 Å) and the Debye length is large (low ionic strength eluents), the pore surface may be considered cylindrical and the relationship between the electrostatic surface potential and the adsorbed IPR can be obtained via the following relationship [5].

$$\Psi^\circ = \frac{z_H[LH]F}{\kappa\varepsilon_0\varepsilon_r}\frac{I_0(\kappa r)}{I_1(\kappa r)} \tag{6.1}$$

where $I_0(\kappa r)$ and $I_1(\kappa r)$ are the modified Bessel functions of the first kind (orders of 0 and 1), r is the pore radius and κ is the inverse Debye length, z_H is the charge of the IPR, $[LH]$ is the surface concentration of the IPR, ε_0 is the electrical permittivity of vacuum, ε_r is the dielectric constant of the mobile phase, and F is the Faraday constant.

Pore size was also found to be the main factor affecting separation selectivity of C18 columns from different manufacturers, compared to evaluate the applicability of sequence-specific retention calculator peptide retention prediction algorithms. Differences in end capping chemistry did not play a major role while the introduction of embedded polar groups to the C18 functionality enhanced the retention of peptides containing hydrophobic amino acid residues with polar groups [6].

Another issue that deserves consideration is the measurement of the surface area of a sorbent after derivatization with alkyl chains. It is generally accepted that the total surface area of a stationary phase after ligand attachment is lower than the surface area of the bare silica because bonded ligands occupy volume inside the pore [7].

Interpretation of results is far from easy. First, when the measurement is performed via the BET method, the surface area of a nitrogen molecule on a hydrophobic surface was found to be higher than its surface area on bare silica [8]. Moreover, it can be postulated that the fact that a fraction of pore volume is occupied does not mean that the surface area is decreased, since alkyl chains expel eluent from pores and if they are elongated, the exposed surface inside the pore can also be higher than before attachment. This can be confirmed by the "phase collapse" that may occur under highly aqueous mobile phases. The hydrophobic staking of adjacent alkyl chains, folded to minimize the area exposed to the aqueous eluent, results in a strong retention loss due to the decrease of surface area; it follows that they are usually elongated.

It should also be noted that nitrogen is not an unbiased probe adsorbate. Obviously the surface accessibility for irregular materials depends on the size of the probe molecule: a large probe cannot follow the irregularity of the surface. Analyte molecules are usually larger than nitrogen molecules, and may not be able to penetrate all pores. Thus only a fraction of the surface area is involved in analyte retention. To fulfill IUPAC recommendations for surface characterization methods [9], the most suitable method depends on the specific application. Recently an approach that involves IPR ion adsorption proved effective. The best probe to determine the packing area accessible to the IPR is the IPR itself. The best estimate of the chromatographic bonded phase packing area is lower than the area found by N_2 sorptometry because this method estimates only the chromatographically accessible surface area [5].

A major issue a chromatographer faces when dealing with most stable silica-based bonded phase columns is the presence of residual silanols whose surface acidity is a very undesirable property. First, they impair lipophilicity determination by liquid chromatography [10]. The ligand bonding density actually shows how well the original silica surface is shielded by the ligand chains. However, after the chemical bonding of the stationary phase, a large percentage of silanol groups remain on the silica surface. They did not react with alkylsilane ligands because of steric hindrance and their residual surface concentration is 8 to 9 μmol/m^2; their pK_a ranges from 5 to 7 and increases to 10 when columns are produced according to hybrid particle technology [11].

The free silanol problem affects even the most highly purified silica supports, including those considered to be the least acidic [12]. Strongly bound metal ions present in the silica inductively enhance the acidity of silanol groups in close proximity and may also coordinate electron donating (Lewis base) analytes, thereby impairing chromatographic performance and silica long-term stability. At eluent pH higher than the pK_a of the silanols, they ionize, thereby providing a negative stationary phase charge density even in the absence of IPRs. If the IPR is negative, the actual surface charge density is higher than that estimated on the basis of its adsorption isotherm; if the IPR is positively charged, the net surface charge concentration is lower than the estimated one since its adsorption must first counterbalance the negative charge.

Moreover. the heterogeneous surface chemistry of the silica–alkyl bonded phase results in poor peak shapes, band deformation, and nefarious losses in column efficiency when basic compounds are analyzed. This is due to the fact that the adsorption equilibrium constant of an analyte on an alkyl ligand site is very low and these sites are very abundant. Conversely, the saturation capacity of free silanols is very low since they are not numerous. However, the adsorption of the analyte on these "active sites" is very strong; the adsorption-desorption kinetics is much slower and molecules captured by these " active" sites are released much later than molecules adsorbed on low-energy alkyl ligand sites. As a result, tailing of the band occurs and, at variance with tailing due to column overload, it does not improve with decreasing the amount of sample injected. The reason is that the "active sites" are always saturated and if the sample amount decreases, the fraction of the analyte molecules interacting with free silanols increases, the band profile deteriorates and retention increases.

Retention of acids can also be negatively affected by ionized free silanols due to electrostatic exclusion phenomena [13]. Free silanols are usually reduced via a procedure known as end capping in which silanol groups are transformed into trimethylsilyl groups via a small silylating agent such as trimethylchlorosilane. Interactions of ionizable analytes with the free silanols remaining after end capping (that still represent a significant percentage of the original number, but most are not accessible) may be reduced by use of mobile phases buffered at low pH, to suppress silanol ionization, or at high pH, to suppress solute ionization.

Several adsorbable quenchers have been examined to suppress the base-attracting and acid-repulsing effects of free silanol in liquid chromatography [14]. Better peak symmetry and system efficiency can be achieved by use of eluents containing a variety of silanol blockers: amines (particularly triethylamine and dimethyloctylamine) or divalent cations are commonly used for this purpose [15], but the great potential of ionic liquids (ILs) as silanol screening agents for separating basic compounds was recently assessed [16–18]. A specific attribute of ILs is their ability to produce strong proton donor–acceptor intermolecular interactions; it follows that they are easily predicted to affect the hydroxy groups of silica supports. ILs were successfully used as silanol suppressors in totally aqueous mobile phases [19]. The addition of ILs reduced the asymmetry factor, thereby improving resolution and separation efficiency. The silanol blocking activity of ILs strongly surpasses that of classical alkylamine additives [16]. Furthermore, ILs may act as IPRs (see Chapter 7) since they simultaneously improve the yield of a separation and change the retention factors.

Use of alkylamide phases, in which alkyl chains are attached to the silica surface via an alkylamide group, reduces interactions with free silanols, by an internal masking mechanism. Residual silanols interact by hydrogen bonding with the embedded amide groups and thus become less active toward analytes. The embedded polar amide groups lessen the hydrophobicity of these phases compared to that of C18 bonded phases prepared from the same silica. Improved peak shapes of ionizable compounds were reported and this stationary phase was successfully used under IPC conditions to analyze streptomycin and its dihydrostreptomycin derivative in food [20]. Similarly an unusual C12 stationary phase with embedded polar urea group successfully achieved IPC baseline separation of seven tetracyclines [21]. The good stability of this phase under highly aqueous conditions is an additional benefit compared to conventional alkyl-bonded phases. It is worth noting that the presence of IPRs in an eluent presents a double advantage: improving oppositely charged analyte retention while improving peak shape.

5.1.2 Other Reversed Phase Stationary Phases

Even if silica-based bonded stationary phases play a major role in IPC separations, other chromatographic packings are gaining popularity because their pH stability is also remarkable. Among them, porous graphitized carbon (PGC) stationary phases exhibit surprising and exclusive properties [22]. PGC is a very insoluble and stable separation medium that exhibits a highly ordered crystalline surface. It consists of well organized benzene rings that provide a very flat surface; it does not need to be functionalized because it is hydrophobic in its own right.

The high density of π–π electrons sets PGC apart among chromatographic packings for resolution of difficult-to-separate positional isomers and diastereoisomers. The presence of surface delocalized electrons singles PGC out for retention of conjugate compounds. The retention patterns of a mixture of analytes may be completely different on alkyl silica and PGC stationary phases. Interestingly, the pH limitations of silica-based reversed phase packings do not hold for PGC and it is possible to use a wide pH range (1 to 13) of the mobile phase. Additionally, the equilibration time for the adsorption of the IPR was faster [23].

Fluorinated IPRs may be harmful for a PGC stationary phase since they may oxidize it, thus reducing its stability at low pH [24]. PGC stationary phases were used in chiral IPC, also at column temperatures below 0°C [25] to enantioseparate closely related amino alcohols [26,27] and also to separate peptides and amino acids with perfluorinated IPRs [22,23,28], and inorganic anions in pharmaceutical compounds [29].

Poly(styrene-divinylbenzene) (PS-DVB) is a well studied stationary phase material; it is hydrophobic and thus does not need to be functionalized. Nevertheless PS-DVB with attached ligands can extend the scope of its application. Its unparalleled pH stability combines with a mechanical resistance that may be sometimes inadequate. Its monolytic form proved superior to columns packed with microparticulate sorbents to separate and identify isomeric oligonucleotide adducts using triethylammonium bicarbonate as the IPR [30].

The role of pore size in non-silica-based supports is similarly significant. In perfusion chromatography, particulate stationary phases are designed to improve the access of molecules to the inner surfaces of the particles. A perfusion stationary phase consisting of highly cross-linked PS-DVB (10 μm particle size), is compatible with back pressures up to 200 bar and resists pH values from 1 to 14 and high ionic strengths and temperatures. Particles through-pores (600 to 800 Å) and diffusive pores (800 to 1500 Å), permit a combination of convective and diffusive transport of the molecule through the column making separations 10 to 100 times faster than in conventional chromatography while preserving the resolution and column loading capacity. This stationary phase was successfully employed to separate maize proteins under the conditions [31].

IPC on non-porous alkylated PS-DVB particles provided high resolution separation of nucleic acids with very short analysis times [32]. Due to its intrinsic pH stability, a PS-DVB monolith was used to bring pH into play for selectivity adjustment in separation of the peptide mixture. Since selectivities were significantly different with acidic and alkaline eluents, a two-dimensional IPC at high pH and at low pH was successfully performed [33].

5.2 NORMAL PHASE IPC AND OTHER STATIONARY PHASES

While a hydrophobic ion-pair is retained on hydrophobic stationary phases better than an ionized analyte, the retention of the duplex on normal phases is easily predicted to be lower than that of the ionized analyte because polar interactions are reduced. Actually the trend of k versus IPR concentration under normal phase IPC is the opposite of reversed phase IPC [34]. An aminopropyl, a cyanoethyl, and a silica stationary phase were compared for the analysis of alcohol denaturants. The cyanoethyl phase was selected and anionic IPRs were used to reduce retention of cationic analyte, suppressing their interactions with negatively charged silanols [34]. The unusual normal phase was also applied in chiral IPC [35] in the IPC of quaternary ammonium compounds [36] and to determine nitrite and nitrate [37]. For the selective enrichment of glycopeptides from glycoprotein digests, charged peptides were paired with oppositely charged sodium chloride that played the atypical role of an IPR. Still more bizarre was the choice to add the IPR to the injection solution; peptide molecules converted into their neutral forms by ion-pairing became more hydrophobic, thus diminishing their ability to bind to the neutral Sepharose column [38].

Oddly, IPC was also performed on a classical ion exchange column to separate 17 anionic, neutral, and cationic arsenic species in a single chromatographic run, thanks to a multiplicity of retention modes on this packing material [39]. Ion pairing proved also valuable in size exclusion chromatography of sulfonated lignins [40].

In Chapter 6, recent developments in column technology will be discussed along with current strategies to obtain fast separation and increased efficiency (breakthroughs of sub-2-μm particles and monoliths).

REFERENCES

1. Szabo, Z. et al. Analysis of nitrate ion in nettle (*Urtica dioica* L.) by ion-pair chromatographic method on a C30 stationary phase. *J. Agr. Food Chem.* 2006, 54, 4082–4086.
2. Yokoyama, Y. et al. Optimum combination of reversed phase column type and mobile phase composition for gradient elution ion-pair chromatography of amino acids. *Anal. Sci.* 1997, 13, 963–967.
3. Krach, C. and Sontag, G. Determination of some heterocyclic aromatic amines in soup cubes by ion-pair chromatography with coulometric electrode array detection. *Anal. Chim. Acta* 2000, 417, 77–83.
4. Weber, S.G. Theoretical and experimental studies of electrostatic effects in reversed phase liquid chromatography. *Talanta* 1989, 36, 99–106.
5. Cecchi, T. Use of lipophilic ion adsorption isotherms to determine the surface area and the monolayer capacity of a chromatographic packing, as well as the thermodynamic equilibrium constant for its adsorption. *J. Chromatogr. A.* 2005, 1072, 201–206.
6. Spicer, V. et al. Sequence-specific retention calculator: family of peptide retention time prediction algorithms in reversed phase HPLC: applicability to various chromatographic conditions and columns. *Anal. Chem.* 2007, 79, 8762–8768.
7. Rustamov, I. et al. Geometry of chemically modified silica. *J. Chromatogr. A.* 2001, 913, 49–63.
8. Giaquinto, A. et al. Surface area of reversed phase HPLC columns *Anal. Chem.* 2008, 80, 6358–6364.
9. Rouquerol, J. et al. Recommendations for characterization of porous solids. *Pure Appl. Chem.* 1994, 66, 1739–1758.
10. Altomare, C. et al. Linear solvation energy relationships in reversed phase liquid chromatography: examination of Deltabond C8 as stationary phase for measuring lipophilicity parameters. *Quant. Struct. Act. Rel.* 2006, 12, 261–268.
11. Gritti, F. and Guiochon, G. Peak shapes of acids and bases under overloaded conditions in reversed phase liquid chromatography, with weakly buffered mobile phases of various pH: a thermodynamic interpretation. *J. Chromatogr. A.* 2009, 1216, 63–78.
12. Gilroy, J.J., Dolan, J.W., and Snyder, L.R. Column selectivity in reversed phase liquid chromatography IV: type B alkyl–silica columns. *J. Chromatogr. A.* 2003, 1000, 757–778.
13. Dai J. and Carr, P.W. Role of ion-pairing in anionic additive effects on the separation of cationic drugs in reversed phase liquid chromatography. *J. Chromatogr. A.* 2005, 1072, 169–184.
14. Righetti, P.G. et al. State of the art of dynamic coatings. *Electrophoresis* 2001, 22, 603–611.
15. Reta, M. and Carr, P.W. Comparative study of divalent metals and amines as silanol blocking agents in reversed phase liquid chromatography *J. Chromatogr. A.* 1999, 855, 121–127.

16. Marszałł, M.P. and Kaliszan, R. Application of ionic liquids in liquid chromatography *Crit. Rev. Anal. Chem.* 2007, 37, 127–140.
17. Ruiz-Angel, M.J., Carda-Broch, S., and Berthod, A. Ionic liquids versus triethylamine as mobile phase additives in the analysis of β blockers, *J. Chromatogr. A.* 2006, 1119, 202–208.
18. Marszałł, M.P., Bączek, T., and Kaliszan, R. Reduction of silanophilic interactions in liquid chromatography with the use of ionic liquids. *Anal. Chim. Acta* 2005, 547, 172–178.
19. He, L. et al. Effect of 1-alkyl-3-methylimidazolium-based ionic liquids as the eluent on the separation of ephedrines by liquid chromatography. *J. Chromatogr. A.* 2003, 1007, 39–45.
20. Viñas, P., Balsalobre, N., and Hernández-Córdoba, M. Liquid chromatography on an amide stationary phase with post-column derivatization and fluorimetric detection for the determination of streptomycin and dihydrostreptomycin in foods. *Talanta* 2007, 72, 808–812.
21. Kallel, L. et al. Optimization of the separation conditions of tetracyclines on a preselected reversed phase column with embedded urea group. *J. Sep. Sci.* 2006, 29, 929–935.
22. Adoubel, A. et al. Separation of underivatized small peptides on a porous graphitic carbon column by ion-pair chromatography and evaporative light scattering detection. *J. Liq. Chromatogr. Rel. Technol.* 2000, 23, 2433–2446.
23. Chaimbault, P. et al. Ion pair chromatography on a porous graphitic carbon stationary phase for the analysis of 20 underivatized protein amino acids. *J. Chromatogr. A.* 2000, 870, 245–254.
24. Rinne, S. et al. Limitations of porous graphitic carbon as stationary phase material in the determination of catecholamines. *J. Chromatogr. A.* 2006, 1119, 285–293.
25. Karlsson, A. and Charron, C. Reversed phase chiral ion-pair chromatography at a column temperature below 0°C using three generations of Hypercarb as solid phase. *J. Chromatogr. A.* 1996, 732, 245–253.
26. Karlsson, A. and Karlsson, O. Chiral ion-pair chromatography on porous graphitized carbon using N-blocked dipeptides as counter ions. *J. Chromatogr. A.* 2001, 905, 329–335.
27. Karlsson, A. and Almgren, K. Reversal of enantiomeric retention order by using a single N-derivatized dipeptide as chiral mobile phase additive and porous graphitised carbon as stationary phase. *Chromatographia* 2007, 66, 349–356.
28. Petritis, K. et al. Parameter optimization for the analysis of underivatized protein amino acids by liquid chromatography and ion spray tandem mass spectrometry. *J. Chromatogr. A.* 2000, 896, 253–263.
29. Okamoto, T., Isozaki, A., and Nagashima, H. Studies on elution conditions for the determination of anions by suppressed ion interaction chromatography using a graphitized carbon column. *J. Chromatogr. A.* 1998, 800, 239–245.
30. Xiong, W. et al. Separation and sequencing of isomeric oligonucleotide adducts using monolithic columns by ion-pair reversed phase nano HPLC coupled to ion trap mass spectrometry. *Anal. Chem.* 2007, 79, 5312–5321.
31. Rodriguez-Nogales, J.M., Garcia, M.C., and Marina, M.L. Development of a perfusion reversed phase high performance liquid chromatography method for the characterisation of maize products using multivariate analysis. *J. Chromatogr. A.* 2006, 1104, 91–99.
32. Dickman, M.J. Post-column nucleic acid intercalation for the fluorescent detection of nucleic acids using ion-pair reverse phase high performance liquid chromatography. *Anal. Biochem.* 2007, 360, 282–287.
33. Toll, H. et al. Separation, detection, and identification of peptides by ion-pair reversed phase high performance liquid chromatography–electrospray ionization mass spectrometry at high and low pH. *J. Chromatogr. A.* 2005, 1079, 274–286.

34. Daunoravicius, Z. et al. Simple and rapid determination of denaturants in alcohol formulations by hydrophilic interaction chromatography. *Chromatographia* 2006, 63, 373–377.
35. Hu, M.H. and Xu, X.Z. Enantiomeric separation of amino alcohols by ion-pair chromatography. *Chin. Chem. Lett.* 2001, 12, 355–356.
36. Bluhm, L.H. and Li, T. *Sci.* Effect of analogue ions in normal phase ion-pair chromatography of quaternary ammonium compounds. *J. Chromatogr. Sci.* 1999, 37, 273–276.
37. Butt, S.B, Riaz, M., and Iqbal, M.Z. Simultaneous determination of nitrite and nitrate by normal phase ion-pair liquid chromatography. *Talanta* 2001, 55, 789–797.
38. Ding, W., Hill, J.J., and Kelly, J. Selective enrichment of glycopeptides from glycoprotein digests using ion-pairing normal phase liquid chromatography *J. Anal. Chem.* 2007, 79, 8891–8899.
39. Kohlmeyer, U. et al. Benefits of high resolution IC-ICP-MS for the routine analysis of inorganic and organic arsenic species in food products of marine and terrestrial origin. *Anal. Bioanal. Chem.* 2003, 377, 6–13.
40. Brudin, S. et al. One-dimensional and two-dimensional liquid chromatography of sulphonated lignins. *J. Chromatogr. A.* 2008, 1201, 196–201.

6 Developments in Column Technology and Fast IPC

Since the infancy of modern high performance liquid chromatography (HPLC) in the late 1960s, users have required improved columns to handle more advanced separations. Over the past decade, the enduring development of HPLC has focused on upgrading of column technology and instrumentation to increase the speed and efficiency of separations. Better columns and instrumentation represent the main needs of analytical laboratories to improve data quality and increase sample throughput. Real-time quality control and field analysis are imperative in certain disciplines and these needs helped raise the level of interest in high speed HPLC. Column researchers and manufacturers responded to these needs by enhancing the efficiency and reliability of packing materials.

In Chapter 5 the peculiarities of different stationary phases materials were discussed, however column technology strongly influenced the upgrade of the IPC technique. Development in fast separations has been primarily related to ultrahigh pressure or ultrahigh performance liquid chromatography (U-HPLC), monolithic columns and high temperatures. In Chapter 12 the role played by high temperature to obtain fast separations will be described in detail. In this chapter, the novel state-of-the-art column technology is illustrated.

6.1 ULTRAHIGH PERFORMANCE LIQUID CHROMATOGRAPHY (U-HPLC)

Theoreticians have known for years that particle size reduction results in more efficient columns; consequently, column performance has been a major research area for decades. The first innovative packings devised in the 1960s were the 40- to 50-μm pellicular (porous-layer bead) types. Their thin porous coatings allowed rapid solute mass transfer into and out of the packing, resulting in better chromatographic efficiency relative to the large (100 μm) and irregularly shaped porous particles used for liquid chromatography separations then. However, their low surface area impaired their sample capacity. Small porous particles (less than 20 μm) were introduced in the 1970s because they combined better efficiency and high capacity. As a natural progression, 5 μm became the standard particle diameter in the late 1980s.

High performance 3-μm particles gained popularity in the early 1990s. For example, a fast (<50 s) IPC determination of nitrite and nitrate on a short 3.0 cm × 0.46 cm 3-μm reversed phase column was validated against conventional ion chromatography and was configured for continuous monitoring of real tap water samples allowing up to 60 analyses per hour, matching the FIA analysis rate [1]. Concurrently the pellicular particles reappeared, also as rather small particles: since the longer diffusive paths are not accessible, band broadening is kept under control and efficiency is improved with lower back pressure compared to the totally porous particles. However, the most promising advance is the breakthrough in the early 2000s of sub-2-μm particles and the commercial releases of several high pressure compatible HPLC platforms [2–4].

The following section demonstrates that short columns packed with sub-2-μm particles and run at high flow rates provide fast (several minutes) and ultrafast (seconds to a few minutes) separations while maintaining or improving resolution.

The efficiency of a chromatographic method can be defined as its ability to elute the same kind of molecule at the same time. During the transit of molecules from the injector along the column length to the detector, band spreading occurs and limits efficiency. The extra column band broadening takes place outside the column (injector, connecting tubing and fittings, column frits, and detector flow cell) and is mainly a function of system configuration.

Band broadening within the chromatographic column is described as a function of mobile phase linear velocity (u) by the well known van Deemter equation [5] that relates the height equivalent to the theoretical plate (HETP) to u: the lower the HETP, the higher the plate number per unit length of a column. It follows that the highest efficiency is obtained for the shortest HETP. The van Deemter equation is:

$$HETP = A + \frac{B}{u} + (C_m + C_s)u \qquad (6.1)$$

The A term is related to multipath effects (eddy diffusion) that are independent of mobile phase flow rate. Analyte molecules can follow multiple pathways of differing lengths that spread the analyte molecules apart and cause peak broadening. Since the smaller the particles, the lower the difference among molecule "walks" along the column, the A term is linearly dependent on particle diameter, d_p, according to λ, a structure factor:

$$A = 2\lambda d_p \qquad (6.2)$$

The B constant is a function of longitudinal diffusion that serves to spread analyte molecules apart and is linearly dependent on D_m, the diffusion coefficient of an analyte in the mobile phase according to an obstruction factor designated γ:

$$B = \frac{2\gamma D_m}{u} \qquad (6.3)$$

The C_m and C_s constants are related, respectively, to the mass transfer resistance between the stationary and mobile phase as well as between the mobile and stationary phase, i.e., to the time needed for the analyte molecules to equilibrate between the mobile and stationary phases. If the resistance is high and the time scale of the equilibrium is long, some of the analyte molecules may not interact with the stationary phase, and are driven by the mobile phase down the column. Other molecules may be left behind because they are slow to detach from the stationary phase. Band broadening occurs in both cases, especially at higher mobile phase flow rates. Smaller stationary phase particles are expected to reduce this equilibration time. When bonded stationary phases or very thin stationary phase films are used, the mass transfer resistance in the stationary phase is relatively small and C_m becomes dominant:

$$C_m = \frac{f(k)d_p^2}{D_m}$$
(6.4)

C_m is proportionally dependent on a function of k (analyte retention factor) and on the particle diameter squared; it can be significantly reduced by reducing particle size. This involves a simultaneous drop of A. In a packed column at high flow rates, the A and C_m terms in the van Deemter equation are more significant than B. When Equation 6.1 is derived as a function of the eluent velocity, its optimum value can be found at the minimum of the van Deemter curve that represents the ideal flow velocity (u_{opt}) where maximum column efficiency (minimum theoretical plate height, maximum theoretical plates per unit length) is obtained [5]:

$$u_{opt} = (B/C)^{1/2} \approx 3D_m/d_p$$
(6.5)

It follows that the drop in particle size results in increased optimum mobile phase velocities. The analyte transfer is faster since its diffusion path length is shorter and the peak remains narrow because the analyte molecules spend less time in the stagnant mobile phase where band broadening occurs. The particle size reduction has another interesting outcome: the van Deemter curves tend to flatten out at higher linear velocities and higher mobile phase flow rates can be safely used without compromising efficiency. This concept is illustrated in Figure 6.1 that shows typical van Deemter curves for 10, 5, and sub-2-μm bonded spherical silica particles. The chromatographic run time for comparable separation decreases from hours to minutes and even seconds. Unfortunately, the pressure drop across a column, ΔP, necessary to give a mobile phase linear velocity, u, is given by

$$\Delta P = \frac{\varphi \eta L}{d_p^2} u$$
(6.6)

where φ is the column flow resistance factor, η is the viscosity of the mobile phase, and L is the column length [6]. Clearly, the main drawback of this strategy is the pressure limitation of conventional HPLC apparatus. With higher flow rates, Joule heating in the column during U-HPLC analysis may also be a concern. The sub-2-μm particles can be packed in capillary or stainless steel columns, provided added flexibility for the procedure. Column diameter is proportional to the amount of sample that can be injected on a column without overloading it while maintaining good peak shape. Smaller diameter columns provide increased sensitivity.

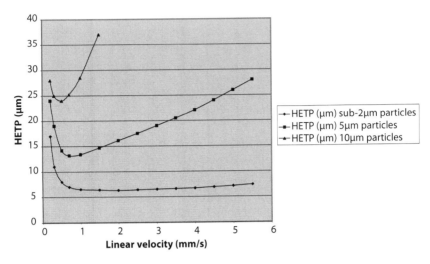

FIGURE 6.1   The van Deemter curves for 10, 5, and sub-2-μm bonded spherical silica particles.

Microcapillary (0.200 ÷ 0.300 mm diameter) and nanocapillary (0.075 ÷ 0.100 mm diameter) columns limit solvent consumption and interface more easily with mass spectrometer detectors. They can assay only small amounts of sample and are superior for managing Joule heating due to their enhanced surface area to volume ratio and lower volumetric flow rate (μL/min) for a given linear mobile phase velocity (mm/sec).

To prevent heating, stainless steel columns are usually microbore (< 2 mm diameter); their features make them suitable for the analysis of larger sample sizes, which in turn improves the detection and quantification of minor components in a mixture. The column length must be drastically reduced to run a chromatographic method on a conventional HPLC apparatus. However, when a modern ultrafast column 15 mm long with an internal diameter of 2.1 mm packed with 1.8-μm particles and a total void volume around 33 μL is used, peaks are often only a few microliters wide. Conventional equipment developed for typical 150 ÷ 250 mm × 4.6 mm analytical columns (total column void volumes = 1.6 ÷ 2.6 mL, respectively) must be upgraded to avoid the extra column band broadening, thus impairing the advantages of these small columns.

Similarly the detector time constant must be adjusted because peaks can be only a second or two wide for short, narrow-bore columns run at high flow rates; a large time constant would make narrow peaks appear artificially broad. If the column length is not drastically reduced, such sub-2-μm packings demand ultrahigh pressure pump systems to overcome the high back pressure (>400 bar).

The required instrument hardware was only recently commercialized; this may explain why the IPC literature concerning U-HPLC is very limited and contemporary. For example, ion-pair U-HPLC coupled with electrospray time-of-flight mass spectrometry played a crucial role in development of rapid and sensitive analytical techniques for the characterization and compositional analysis at the disaccharide level of heparin and heparan sulfate. Standards were eluted in less than 5 min [7]. The same technique was found valuable for structural analysis of heparin and heparan sulfate; the heterogeneity of preparative-scale size exclusion chromatography fractions was examined using the volatile tributylamine as the IPR [8].

We submit that this research field is not mature and future developments of the IPC technique will focus on U-HPLC because the peak capacity of U-HPLC is impressive. Separations can be accomplished ten times faster while maintaining or improving resolution and sensitivity. We can conclude that sub-2-μm particles will make a new realm of chromatography reachable. The combination of increased peak capacity (due to higher efficiency) and speed of analysis (due to higher flow rates) has improved sensitivity (due to less band broadening and hence sharper peaks), allowing chromatographic runs to be downsized. Of course, further reductions in porous particle size to achieve even faster separations will require auxiliary implementation of instrument hardware.

6.2 MONOLITHIC COLUMNS

In the past decade, extensive attention has been paid to the novel porous monoliths that have been applied gradually in microscale chromatographic separation techniques such as capillary liquid chromatography [9]. Monoliths can be described as continuous porous separation media because they consist of single particle polymer pieces without interparticular voids. The use of monolithic columns pursues the same goal as U-HPLC separation. Chromatographers using both methods seek fast separations without compromising efficiency and resolution. Both approaches smooth the progress of application to mass spectrometers, but monoliths introduce considerable advantages over particulate columns and represent attractive alternatives for micro- and nanoscale separations.

High permeability due to high porosity and small skeleton size are the distinguishing physical features of monoliths. Their open channel structures allow a chromatographer to use higher flow rates while maintaining moderate back pressure on relatively long columns using conventional equipment. Fast separations are highly efficient due to the large number of theoretical plates per unit pressure drop. The mobile phases are forced through the porous monolithic media and the resulting convective flow [10] increases the mass transfer rate because it prevents the formation of a stagnant mobile phase where detrimental band broadening occurs. Additional advantages include versatile surface modification and good peak capacity. Fritless designs based on the covalent immobilization of the material at the inner wall of the capillary avoid frit-related problems.

The porous structure is due to the presence of suitable inert diluents (porogens) during the polymerization process [11]. To obtain optimal chromatographic properties, a number of variables must be optimized: polymerization temperature, composition of porogen and amount of cross-linking of monomer and initiator were found to be the key parameters to control average pore size and overall surface area [11]. Morphology and porosity of monoliths are usually studied by scanning electron microscopy, mercury intrusion porosimetry, and inverse size exclusion chromatography.

During the past 10 years, in addition to silica-based monoliths [12], a broad range of organic polymeric monoliths has been studied. Their most advantageous attribute is their chemical stability over a wide pH range. The most common organic monoliths were the results of methacrylate [13] and styrene [14] monomers. Some examples that confirm the utility of monolithic columns in IPC are described below.

A monolithic silica-based C18 stationary phase was used under high flow rate condition (2 mL/min) without significant back pressure in IPC analysis of a recently discovered new drug candidate for the treatment of Alzheimer's disease [15]. Nanoscale IPC using a monolithic poly(styrene-divinylbenzene) (PS-DVB) nanocolumn coupled to nanoelectrospray ionization mass spectrometry (nano-ESI-MS) was evaluated to separate and identify isomeric oligonucleotide adducts. Triethylammonium bicarbonate was used as the IPR. Interestingly, the performance of the polymeric monolithic PS-DVB stationary phase significantly surpassed that of columns packed with the microparticulate sorbents C18 or PS-DVB [16].

It is important to emphasize that column downscaling involves easy and efficient coupling to mass spectrometry with improved detection limits because no sample loss occurs. [17]. Due to its intrinsic pH stability, a PS-DVB monolith was used to bring pH into play related to selectivity adjustment in separation of a peptide mixture [18]. It proved to be the right choice also for the direct molecular typing of genes [19] and for the analysis of ribonucleic acids [20].

Recently an aromatic capillary monolithic acrylate material was prepared via a thermally initiated free radical polymerization in the confines of 200-μm inner diameter fused silica capillaries. The material proved to be very robust and was applied to IPC of oligodeoxynucleotides with good reproducibility [21]. Similarly a derivatized poly[(trimethylsilyl-4-methylstyrene)-co-bis(4-vinylbenzyl)dimethylsilane] capillary monolith was used for the same purpose [22].

A comparison of monolithic conventional size, microbore, and capillary poly(p-methylstyrene-co-1,2-bis(p-vinylphenyl)ethane) columns confirmed that the efficiency for analysing proteins and oligonucleotides improved with decreasing column internal diameter, even if monolithic capillary columns up to 0.53 mm internal diameter were successfully used for the fractionation of the whole spectrum of biopolymers including proteins, peptides, and oligonucleotides as well as double-stranded DNA fragments under IPC conditions [14,23].

A monolithic column provided an ultra-fast (15-s) IPC separation of common inorganic anions using tetrabutylammonium-phthalate as the IPR and was successfully validated against standard IC [24]. Monolithic capillaries in IPC also brought increased sensitivity to the analysis of very low quantities of RNAs [25]. We conclude that monoliths represent a promising approach for efficient and fast separations but require further investigation.

REFERENCES

1. Connolly, D. and Paull, B. Rapid determination of nitrite and nitrate in drinking water samples using ion-interaction chromatography. *Anal. Chim. Acta* 2001, 441, 53–62.
2. Chesnut, S.M. and Salisbury, J.J. The role of UHPLC in pharmaceutical development. *J. Sep. Sci.* 2007, 30, 1183–1190.

3. Jacob, S.S., Smith, N.W., and Legido-Quigley, C. Assessment of Chinese medicinal herb metabolite profiles by UPLC-MS-based methodology for the detection of aristolochic acids. *J. Sep. Sci.* 2007, 30, 1200–1206.

4. Walles, M. et al, Comparison of sub-2-μ particle columns for fast metabolite ID *J. Sep. Sci.* 2007, 30, 1191–1199.

5. Snyder, L.R. and Kirkland, J.J. *Introduction to Modern Liquid Chromatography.* Wiley: New York, 1979.

6. Giddings, J.C., *Unified Separation Science.* Wiley: New York, 1991, p. 65.

7. Korir, A.K. et al. Ultraperformance ion-pair liquid chromatography coupled to electrospray time-of-flight mass spectrometry for compositional profiling and quantification of heparin and heparan sulfate. *Anal. Chem.* 2008, 80, 1297–1306.

8. Eldridge, S.L. et al. Heterogeneity of depolymerized heparin SEC fractions: to pool or not to pool? *Carbohydr. Res.* 2008, 343, 2963–2970.

9. Wu, R. et al. Recent development of monolithic stationary phases with emphasis on microscale chromatographic separation. *J. Chromatogr. A.* 2008, 1184, 369–392.

10. Liapis, A.I., Meyers, J.J., and Crosser, O.K. Modeling and simulation of the dynamic behavior of monoliths. *J. Chromatogr. A.* 1999, 865, 13–25.

11. Svec, F. and Fréchet, J.M. Temperature: a simple and efficient tool for the control of pore size distribution in macroporous polymers. *Macromolecules* 1995, 28, 7580–7582.

12. Motokawa, M. et al. Monolithic silica columns with various skeleton sizes and through-pore sizes for capillary liquid chromatography. *J. Chromatogr. A.* 2002, 961, 53–63.

13. Lee, D., Svec, F., and Fréchet, J.M. Photopolymerized monolithic capillary columns for rapid micro high-performance liquid chromatographic separation of proteins. *J. Chromatogr. A.* 2004, 1051, 53–60.

14. Trojer, L. et al. Monolithic poly(p-methylstyrene-co-1,2-bis(p-vinylphenyl)ethane) capillary columns as novel styrene stationary phases for biopolymer separation. *J. Chromatogr. A.* 2006, 1117, 56–66.

15. Mancini, F. et al. Monolithic stationary phase coupled with coulometric detection: development of an ion-pair HPLC method for the analysis of quinone-bearing compounds. *J. Sep. Sci.* 2007, 30, 2935–2942.

16. Xiong, W. et al. Separation and sequencing of isomeric oligonucleotide adducts using monolithic columns by ion-pair reversed-phase nano-HPLC coupled to ion trap mass spectrometry. *Anal. Chem.* 2007, 79, 5312–5321.

17. Hochleitner, E.O. et al. Analysis of isolectins on non-porous particles and monolithic polystyrene divinylbenzene-based stationary phases and electrospray ionization mass spectrometry. *Int. J. Mass Spectrom.* 2003, 223–224, 519–526.

18. Toll, H. et al. Separation, detection, and identification of peptides by ion-pair reversed-phase high-performance liquid chromatography–electrospray ionization mass spectrometry at high and low pH. *J. Chromatogr. A.* 2005, 1079, 274–286.

19. Oberacher, H. et al. Direct molecular haplotyping of multiple polymorphisms within exon 4 of the human catechol-O-methyltransferase gene by liquid chromatography–electrospray ionization time-of-flight mass spectrometry. *Anal. Bioanal. Chem.* 2006, 386, 83–91.

20. Holzl, G. et al. Analysis of biological and synthetic ribonucleic acids by liquid chromatography-mass spectrometry using monolithic capillary columns. *Anal. Chem.* 2005, 77, 673–680.

21. Bisjak, C.P. et al. Novel monolithic poly(phenyl acrylate-co-1,4-phenylene diacrylate) capillary columns for biopolymer chromatography. *J. Chromatogr. A.* 2007, 1147, 46–52.

22. Jakschitz, T.A.E. et al. Monolithic poly[(trimethylsilyl-4-methylstyrene)-co- bis(4-vinylbenzyl)dimethylsilane] stationary phases for the fast separation of proteins and oligonucleotides. *J. Chromatogr. A.* 2007, 1147, 53–58.

23. Trojer, L. et al. Comparison between monolithic conventional size, microbore and capillary poly(p-methylstyrene-co-1,2-bis(p-vinylphenyl)ethane) high-performance liquid chromatography columns: synthesis, application, long-term stability and reproducibility. *J. Chromatogr. A.* 2007, 1146, 216–224.
24. Hatsis, P. and Lucy, C.A. Ultra-fast HPLC separation of common anions using a monolithic stationary phase. *Analyst* 2002, 127, 451–454.
25. Dickman, M.J. and Hornby, D.P. Enrichment and analysis of RNA centered on ion-pair reversed phase methodology. *RNA* 2006, 12, 691–696.

7 Ion-Pair Reagents (IPR)

The development of most ion-pairing chromatography (IPC) methods started with optimization of the mobile phase composition after an appropriate column was selected. The theory described in Chapter 3 assists the chromatographer to perform educated guesses. Chapters 7 through 10 discuss the main qualitative and quantitative attributes of tunable mobile phase parameters to illustrate their influence on the global performance of the method.

Key factors in IPC separations are the kind and concentration of the ion-pairing reagent (IPR). The popularity of different classes of IPRs changes quickly and this in turn led to dynamic development of new IPC features. As a result, this versatile technique proved capable of tracking novel analytical needs. Figure 7.1 details the distribution of traditional and innovative IPRs in 2008. This chapter explains their attributes and uses. Clearly the major role for positively charged IPRs is still played by classical organic ammonium ions, whereas for negatively charged IPRs, perfluorinated carboxylic acids and chaotropic lipophilic salts are outperforming traditional sulfonium organic ions. The breakthrough of ionic liquids is also noteworthy and the simultaneous use of both positive and negative IPRs gained some interest. The essential role played by IPRs in IPC is described in Chapter 3.

7.1 TRADITIONAL IPRs

Conventional IPC separations were achieved by adding organic amines and ammonium salts of varying chain lengths as cationic IPRs for anionic analytes and alkyl or aryl sulfonates and sulfates as anionic IPRs for cationic solutes. Since these IPRs exhibit good tensioactive properties, IPC was originally named "soap chromatography" [1].

The molecular architecture and the chain length of the IPR greatly affect analyte retention enhancement. At constant eluent composition, log k of an analyte oppositely charged to the IPR raises linearly with the number of carbon atoms in the chain of the alkyl sulfonate [2,3]. The shorter the IPR side chain, the lower its adsorption onto the stationary phase. This involves a lower potential difference between the electrified stationary phase and bulk eluent that in turn results in a lower driving force for analyte retention [4–7]. To attain comparable retention, the effect played by a higher concentration of a less hydrophobic IPR is tantamount to that of a lower concentration of a more lipophilic IPR [8–10]. However, the relationship between adsorbophilicity and chain length of the IPR is only part of the story.

Interestingly, a slightly different retention of a probe analyte can be obtained when the stationary phase concentrations of IPRs of different chain lengths are the same [11]. This phenomenon cannot be explained by electrostatic theories because the potential established at the stationary phase is only a function of the surface

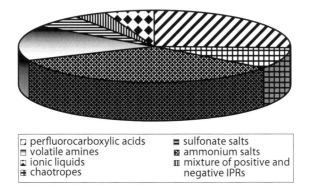

□ perfluorocarboxylic acids	▤ sulfonate salts
▥ volatile amines	⊠ ammonium salts
▧ ionic liquids	▥ mixture of positive and
⊞ chaotropes	negative IPRs

FIGURE 7.1 Extent of use of traditional and novel IPRs in 2008.

charge that in turn depends on IPR surface concentration and is independent of its chain length. At variance with electrostatic theories, the extended thermodynamic approach to IPC quantitatively rationalizes this experimental evidence [4]. The same stationary phase concentrations of IPRs of different chain lengths are related to different IPR eluent concentrations, according to the specific adsorption isotherm of each IPR. This phenomenon can be observed in Figure 7.2 that outlines the adsorption isotherms of two hypothetical IPRs of different chain lengths under the same experimental conditions. The higher the IPR hydrophobicity, the lower the eluent concentration necessary to establish a particular surface charge. For example, to attain a surface in excess of 1.0 μmol/m^2, mobile phase concentrations of 25 and 100 mM, respectively, are necessary for the two IPRs. If the complex formation in

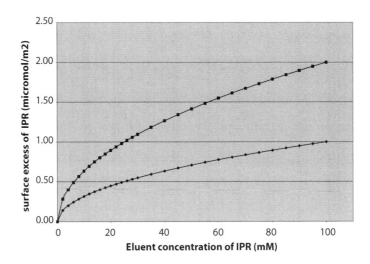

FIGURE 7.2 Adsorption isotherms of two hypothetical IPRs of different hydrophobicities under the same experimental conditions.

the mobile phase is considered, different retentions of probe analytes at constant stationary phase concentrations of IPRs of different hydrophobicities can be easily explained because ion-pairing in the mobile phase occurs to an extent that is a function of the IPR eluent concentrations. On the converse, electrostatic theories cannot explain this phenomenon since they do not take ion-pairing into account.

As to alkylammonium IPRs, it is clear that the retention enhancement of oppositely charged analytes increases with increasing numbers of alkyl nitrogen substituents: monoalkylammonium < dialkylammonium < trialkylammonium < tetraalkylammonium. Again, the longer the alkyl chain, the more effective the IPR [12].

Obviously if the analyte and IPR have the same charge status, the more hydrophobic the IPR, the larger the analyte retention decrease. In this case, the surface charge and the potential of the electrified interface that run counter to analyte adsorption are stronger and the analyte is electrically excluded from the stationary phase.

In summary, the chain length of the IPR and its concentration show very similar effects on retention and their effects can be modulated by lower (higher) concentrations of an IPR when it is strongly (weakly) hydrophobic. Classical IPRs are commonly used in a concentration range between 0.1 mM and 100 mM. As a borderline case, for very hydrophobic IPRs, a permanent coating of the stationary phase does occur and the retention enhancement of oppositely charged analytes is effective even when the IPR presence in the eluent is discontinued. This translates into a cost-effective technique since the IPR can be eliminated from the eluent or only a minimal quantity used because the reversed phase column behaves as an ion exchanger. An advantage of this approach is that the column exchange capacity can easily be varied by adjusting the organic modifier content of the IPR coating solution, which, in turn, influences the amount of adsorbed IPR. Permanent coating of the stationary phase presents the additional advantage that eluted analytes are free from the ionic modifier. However, the stability of the coating must be checked periodically. This strategy was found valuable for fast separation of common anions [13] and in the gradient separation of the entire lanthanide series in a fission product mixture on a reverse phase column coated with one of the best extractants available for lanthanides: di-(2-ethylhexyl) phosphoric acid [14].

7.2   VOLATILE IPRs

Liquid chromatography mass spectrometry (LC-MS) is now routinely used in analytical laboratories. Traditional IPRs are non-volatile salts that are not compatible with MS techniques because they play a major role in source pollution that is responsible for reduced signals. Moreover the final number of charged ions that reach the detector is impaired by ion-pair formation; actually IPRs added to the mobile phase to improve analytes retention exert a profound effect on analyte ionization. Chromatographers who perform IPC-MS must optimize the eluent composition based on both chromatographic separation and compatibility with online detection requirements.

Different strategies devised for effective IPC-MS methods are discussed in greater detail in Chapter 14. This chapter examines the advantages of using volatile

IPRs that hold much promise in this field and also for easy IPC evaporative light scattering detector techniques.

Small organic (e.g., formic and acetic) acids are effective volatile IPRs. They impact the retention behaviors of pH-sensitive compounds, changing their charge status and providing pairing anions that may easily interact with protonated solutes. Many chromatographic separations benefit in terms of retention, resolution, and peak shape under acidic conditions due to suppression of silanol activity. Furthermore, the acidity of these IPRs facilitates the formation of the protonated molecular ion $[M + H]^+$ measured by mass spectrometry in the usual positive ion mode.

However, the use of several homologues of perfluorinated carboxylic acids [7,10,15–48] has increased greatly for peptide, amino acid, hydrophilic metabolite, and ionogenic base applications. They proved to be excellent IPRs even for underivatized amino acids. Again, their presence in the eluent affects retention, lowering the pH of the mobile phase [49,50] and reducing silanol interactions. Trifluoroacetic acid (TFA) is particularly valued for its high water solubility and transparency at 220 nm. Heptafluorobutyric acid [22,23,25,26,38–41] (HFBA) was used for successful coupling to MS and to obtain additional selectivity compared to TFA or pentafluoropropanoic acid (PFPA).

The volatility of perfluorinated IPR was also beneficial when an evaporative light scattering detector was employed [5,17,24,50]. Perfluorinated carboxylic acids proved suitable for preparative chromatography [51]. Interestingly, since they prevent strong IPR build-up in the column, they are particularly useful when gradient elution is performed to reduce analysis time without compromising resolution [28,52–54].

When analytes are sensitive to acids or do not show optimal resolution at low pH, more neutral conditions may be necessary. The employment of small organic acids is not beneficial. In these cases, volatile salts such as ammonium formate and ammonium acetate are the additives of choice [55–62]. The pH of these salt solutions allows both positive and negative ion mode MS detection. The main concern is the limited solubility of the salts in organic solvents, particularly in acetonitrile. Another issue is the pH increase during an organic modifier gradient that usually worsens separation. Easy results with MS were also accomplished with volatile alkyl-,dialkyl-, trialkylamines [25,49,55,56,63–83] often in the formate or acetate forms that yield a good trade-off between retention and selectivity on one hand and the MS response on the other. Interestingly the volatilities of tetrabutyl ammonium acetate [84] and bromide [85] were sufficient for IPC-ESI-MS.

7.3 CHAOTROPIC SALTS

Chaotropic salts are not strongly hydrated; they are known to disrupt the hydrogen bonding structure of water and increase disorder ("chaos") [86]. They are arranged in the Hofmeister series, in which the rank (chaotropicity) of an ion is related to its charge delocalization and polarizability. In order of increasing chaotropicity we have the following series [87–88]: $H_2PO_4^- < HCOO^- < CH_3SO_3^- < Cl^- < NO_3^- < CF_3COO^- < BF_4^- < ClO_4^- < PF_6^-$.

At variance with traditional IPRs that tend to stick very strongly to the stationary phase and to impair the initial column properties also when their presence in the eluent is discontinued, chaotropic IPRs are quite hydrophilic and can be easily removed from the column.

Chapters 2 and 3 discussed the theoretical description of the chaotropic effect. In this chapter we detail the attributes of chaotropic salts as novel IPRs. Among newly introduced IPRs, chaotropic salts achieved good recognition among chromatographers during the past decade, and proved particularly useful in the IPC of basic compounds in low pH eluents [87,89–98]. Basic compounds attracted significant interest based on an estimate that about 80% of drugs include basic functional groups [99]. HPLC separations of basic specimens are challenging based on extreme retention changes as a function of mobile phase pH. Moreover, peak tailing due to silanophilic interactions is a common drawback that causes poor peak efficiency.

The hydrophobicity of the basic analyte has an operative role determining the extent to which the concentrations of chaotropic anions will enhance its retention: the lower the analyte hydrophobicity, the lower the effect of the increase of chaotropic anion concentration [87,100]. Protonated amine retention is stronger as more chaotropic counter ions are used as IPRs [101]. These effects were explained as the results of the dominance of ion-pair interactions in solution; chaotropes dehydrate more easily than kosmotropes. Strong ion-pair interactions entail the exclusion of water molecules between the pair charged species; hence chaotropic anions can yield a neutral complex that is better retained at the hydrophobic stationary phase. The same view was shared by Gritti and Guiochon [102–104] and by Flieger [87] and explains the superiority of the perchlorate ion over dihydrogen phosphate for the IPC of β-blockers [89] and over trifluoroacetate for the IPC of peptides [105].

However, Kazakevich and co-workers demonstrated the importance of the adsorption of chaotropic ions onto the reversed stationary phase [106]. The rank of an ion in the Hofmeister series is another measure of its tendency to accumulate at the stationary phase in RP-HPLC and be quantified via its adsorption isotherm [88]. Clearly a specific surface excess of the chaotropic reagent results in the development of a potential difference between the stationary phase and the bulk eluent, modulating retention of analytes [107]. We discussed in Chapter 3 the way a comprehensive theory can take this experimental evidence into account.

The effect of the nature of a chaotropic mobile phase additive on retention behavior of basic drugs analytes was studied. They influence the selectivity, efficiency, and repeatability [87,95,108]. Interestingly, weakly chaotropic ions such as those coming from acetic [94,109–111] or phosphoric acid were advantageously used as IPRs [16,89–91] and the results may be rationalized by the theory of chaotropicity [95].

Even if anionic chaotropes are the most popular neoteric IPRs, polarizable cations such as sulfonium and phosphonium reagents showed single selectivity toward polarizable anions; their behavior was rationalized on the basis of their chaotropicity. Probe anion retention generally increases in the order of tributylsulfonium < tetrabutylammonium < tetrabutylphosphonium. Interestingly, retention was found to be influenced by the kosmotropic/chaotropic character of both the IPR and the probe anion [93] and this confirms the peculiarities of hydrophobic ion-pairing. Quaternary phosphonium salts provided increased selectivity compared to ammonium in the IPC of heavy metal complexes of unithiol [112].

TABLE 7.1

Dependence of Retention of Positively Charged Analytes on Eluent Concentration of NaClO$_4$

NaCl (mM)	NaClO$_4$ (mM)	k′ Atenolol	k′ Octopamine	k′ 3-Hydroxytyramine	k′ N,N′-dimethyl-benzylamine
0	0	0.23	0.05	0.08	0.30
0	10	0.41	0.13	0.18	0.59
0	25	0.50	0.16	0.21	0.73
0	50	0.58	0.18	0.24	0.86
0	100	0.65	0.19	0.26	1.01
0	500	0.85	0.21	0.31	1.44
500	0	0.52	0.16	0.21	0.77

HPLC: Agilent 1200 equipped with diode array detector (DAD). Analytical column: Zorbax Eclipse XDB-C18, 4.6×75 mm, 3.5 μ. Mobile phase: 30:70 (v/v) mixture of methanol:water acidified with HCl (pH 2.4) and different amounts of NaClO$_4$ and NaCl. Temperature: 20.0 ± 0.1°C. Flow rate: 1.000 ml/min.

For the adsorption of an organic cation, it was demonstrated that retention depends on the hydrophobicity of the anion, according to the following order that reflects well the global hydrophobicities of the ions: HOOCC$_6$H$_4$COO$^-$ ≤ Cl$^-$ ≤ CH$_3$COO$^-$ ≤ HOOCC$_2$H$_4$COO$^-$ ≈H$_2$PO$_4^-$ [103]. Notably, the classical distinction between IPRs and indifferent electrolytes is fading because the latter may mimic the role played by the former to an extent that depends on their chaotropicities. Even Cl$^-$ was demonstrated to be an effective retention enhancer of poorly retained cationic analytes.

Table 7.1 illustrates the effects of increasing concentrations of a typical chaotropic IPR (perchlorate) on the retention of cationic analytes in the low pH region. It is evident that retention increases in a large concentration range of IPR; retention maxima that are often observed at high concentrations of traditional IPRs are missing. This can be explained taking into account that a chaotropic salt has no critical micellar concentration. Since its hydrophobicity is low, adsorption competitions for available ligand sites on the stationary phase are easily predicted to be negligible. NaCl (less chaotropic than NaClO$_4$) is less effective as a retention enhancer when the eluent concentration is the same.

7.4   IONIC LIQUID-BASED IPRs

Ionic liquids (ILs) have been recognized since the early 1900s. They are molten salts with low melting points, usually below 100°C. They were introduced in organic synthesis as a new class of polar, non-molecular solvents in the late 1990s and have recently attracted interest in a variety of fields since solvents are at the heart of most chemical processes and ionic liquids show an exclusive and fascinating wide range of physico-chemical properties.

High thermal stability, negligible vapor pressure, low flammability, good conductivity, wide electrochemical windows, liquid ranges of several hundred of degrees liquid, and chemical stability are the most valuable features of this class of compounds [113]. Their adjustable polarities, solvating abilities, and tunable hard and soft characters distinguish them from classical organic solvents and make possible the development of novel technologies and applications. Another feature is that their ionic liquid environments are very different from those of polar and non-polar organic solvents. Some properties such as thermal stability and miscibility are related to anion nature, while others such as surface tension, viscosity, and density depend on the length and/or shape of the alkyl chain in the cation. However, the low melting points have little to do with fundamental properties of the salts [114,115].

Separation science focuses on room temperature ionic liquids (RTILs), salts that are liquid at ambient temperature. They have been studied as extracting solvents, stationary and mobile phases, mobile phase additives, and other uses. Common RTILs consist of a bulky nitrogen- or phosphorus-containing organic cation (pyridinium or pyrrolidinium, alkyl-imidazolium, ammonium or phosphonium) and a variety of organic and inorganic anions (triflate, dicyanamide, trifluoroacetate, acetate trifluoromethylsulfate, nitrate, perchlorate, bromide, chloride, chloroaluminate, tetrafluoroborate, hexafluorophosphate).

In liquid chromatography and electrophoretic methods, ILs are mostly used in diluted form in aqueous solutions. If its concentration is lowered to the millimolar range, an IL may be used as a mobile phase ionic additive. Their breakthrough for use in RP-HPLC was due to their ability to suppress deleterious effects of silanophilic interactions that represent the main drawback of silica-based stationary phases; they also exhibit many other favorable physical attributes [116].

Peak tailing, band broadening, low plate numbers, and lengthy chromatographic runs are common problems caused by free silanols (see Chapter 5 for more details). Chromatographers have continuously tested numerous adsorbable quenchers to suppress the base-attracting and acid-repulsing effects of free silanols in liquid chromatography. Amines and divalent cations are common silanol-blocking agents [117]. The great potential of ILs as silanol screening agents for the separations of basic compounds was only recently assessed.

The specific attributes of these compounds that provide strong proton donor–acceptor intermolecular interactions were easily predicted to affect the hydroxy groups of silica supports. ILs were successfully used as silanol suppressors and proved superior to classical alkylamine silanol screening reagents. The addition of ILs to aqueous mobile phases reduced the asymmetry factor and improved resolution and separation efficiency. Better peak shapes were associated with changed retention factors. 118–123] and it was soon realized that it is possible to "play" with analyte retention by choosing different ILs and varying their concentrations [119,122]; it follows that they can function as both silanol screening reagents and as innovative and unique IPRs since hydrophobic and ion-pairing interactions with charged analytes are always present in a chromatographic system.

The IPC theory [10] gives a clear and easy rationalization of analyte retention dependence on the presence of ILs used as additives in the mobile phase in low concentrations. In this case, their specific properties as non-molecular polar solvents are not important. In fact, they are salts characterized by a dual nature because of the different properties of the cation and anion. Both contribute to solute retention and improvement of peak shape. IL anions can adsorb on a hydrophobic stationary phase; the amount depend on their positions in the Hofmeister anion series. Similarly, IL cations also can adsorb on C18 stationary phases following a lyotropic series directly related to alkyl chain lengths [108,124,125].

A stationary phase, modified by sorption of IL anions and cations, will show an electrostatic potential difference compared to the bulk mobile phase. The sign of the potential depends on the IL ion that shows the strongest adsorbophilic attitude toward the PR material; for example, if the cation is more adsorbophilic than the anion, the electrified surface will bear a positive charge. In this case, retention is subordinated to ion interactions similar to those of a classical IPC system that makes use of cationic IPRs.

Let us focus on positively charged basic compounds in low pH mobile phases. In the presence of ILs in the mobile phase, these analytes are retained by a combination of electrical and hydrophobic interactions with the stationary phase and with the ions in the mobile phase. The peak shape depends on the kinetics of the interaction. In aqueous mobile phases, electrostatic interactions are usually stronger and slower than hydrophobic ones and lead to band broadening. It can be speculated that the retention changes represent a trade-off between the retention increase due to ion-pairing interactions of the analyte cations and the IL anions and the retention decrease due to screening of detrimental attractive silanophilic interactions by IL cations adsorbed onto the silica-based stationary phase. If the latter effect overcomes the former, an experimental bare analyte retention decrease is observed upon addition of IL additives. This view is supported by the effect of the IL's cation hydrophobicity. The increase of its alkyl chain length increases the screening effects due to stronger adsorption of IL cations onto the hydrophobic stationary phase; thus better silanol screening activity is obtained.

The influence of the nature of the IL anion confirms the importance of ion-pairing for fine tuning retention. The lower the hydrophobicity and ion-pairing ability of the IL anions, the lower the retention. Ion pairing of the cationic analyte and the IL anion decreases silanophilic interactions and this in turn results in better peak shapes, and may eventually increase analyte retention. Actually, even if most research related the retention decrease upon IL addition in the mobile phase to decreased silanophilic interactions, it must be noted that positively charged analytes and IL cations also undergo repulsive electrostatic interactions.

To deconvolve the silanophilic effect from the electrostatic repulsion, a non-silica-based stationary phase may be suitable in research work. On a polystyrene–divinylbenzene reversed phase column, an ethylammonium formate RTIL was not able to produce effective ion-pairing interactions with acidic and basic model compounds, and baseline resolution was only obtained in the presence of classical IPRs (tetrabutylammonium and dodecylsulfate ions, respectively). However, the RTIL was able to mimic the methanol role [123,126]. In summary, IL cations reduce positively charged analyte retention since they (1) screen free silanols and (2) electrify the stationary phase with a positive surface charge that is repulsive for cationic analytes. The hydrophobic character of IL anions is responsible for possible analyte retention increases via ion-pairing.

The mixed mechanism outlined above can be elucidated by analyzing the effects of different salts at the same concentration on the retention of a cationic analyte. The replacement of NaCl by NaBF$_4$ increases retention and slightly improves peak shapes. The BF$_4^-$ anions adsorb on the C18 stationary phase more than Na$^+$, cations thereby providing a negatively charged electrified interphase. BF$_4^-$ associates with the cationic analyte, decreasing its polar character; moreover the potential difference between the stationary phase and the bulk eluent contribute to increase the analyte retention factor.

If NaCl is replaced by the butyl-3-methyl imidazolium (BMIM) chloride IL, a 30% decrease in retention factor associated with a remarkable peak shape improvement is observed. In this case, the IL cation adsorbs on the C18 stationary phase more than Cl$^-$, thereby preventing detrimental attractive silanophilic interaction of the cationic additive. Charge–charge repulsion occurs, the retention factor is lower, and the peak shape is better. The analyte cation is largely retained by hydrophobic fast interactions. When BMIM BF$_4$ IL replaces NaCl, both the cation and anion of the IL adsorb on the C18 surface and all the interactions cited above take place simultaneously and contradict each other. Global retention depends on the extent to which one interaction is stronger than the other [124].

It follows that by increasing the IL concentration in the eluent, overall retention of the analyte may potentially decrease [127], increase [124,128], or remain almost constant if the contradicting effects of the IL cation and anion balance each other [122]; the effect depends on the specific IL content in the mobile phase [129]. The dependence of analyte retention on IL eluent concentration for the increase situation is shown in Figure 7.3. It is worth noting that the shape of the plot is very common in IPC, thereby confirming the ion-pairing aptitude of ILs. Furthermore a reversal of elution sequence with increasing IL concentration was reported [130]. The synergistic contributions of both cationic and anionic

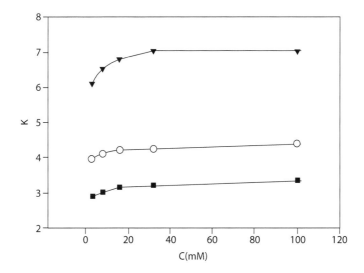

FIGURE 7.3 Effect of 1-ethyl-3-methylimidazolium tertafluoroborate concentrations on retention factors (k′) of basic compounds. Symbols: ● = octopamine; ○ = synephrine; ▼ = tyramine. (From Tang, F. et al. *J. Chromatogr. A.* 2006, 1125, 182–188. With permission from Elsevier Science.)

components of ILs generate their unique properties as mobile phase additives. Experimental works with a wide variety of ILs [131] confirm that the IPC theory can rationalize the analyte behavior in the presence of ILs as mobile phase additives [122,128,132].

Finally, we must emphasize that ILs are certainly not the additives of choice for IPC-MS applications because their non-volatility will create condensation and pollution in the ionization sources. Moreover, slight UV absorbance may limit the use of UV detectors.

7.5   UNUSUAL IPRs

Uncommon IPRs were tested recently. Polymerized acyl monoglycinate surfactant was found to be as effective as sodium dodecylsulfate for the resolution of organic amines [126]. For the analysis of pyridine-based vitamins in infant formulas, dioctylsulfosuccinate produced a unique retention pattern [133]. Among bizarre IPRs, tris(hydroxymethyl)aminomethane was used for the determination of cyclamate in foods. It was selected over different ion-pair reagents such as triethylamine and dibutylamine, based on sensitivity and time economies [134]. Hexamethonium bromide, a divalent IPR, was used successfully to separate sulfonates and carboxylates [135].

Interestingly a crystal violet dye was used as a IPR for indirect photometric detection. Common inorganic anions were separated and detected at the absorption maximum of the dye with classical reversed phase columns. The affinity of the analytes decreased in the order of $S_2O_3^{-2} > SO_4^{-2} > I^- > NO_3^- > Br^- > NO_2^- > Cl^-$, which is the same as that observed with classical anion exchange chromatography [136]. 3,5-Dinitrobenzoic acid proved capable of improving reversed phase retentions of chemical warfare agent derivatives [137] and [S-(R,R)]-(-)-bis(-α-methylbenzyl) amine hydrochloride was an effective IPR for zwitterionic analytes [138].

An atypical simultaneous presence of both anionic and cationic modifiers was reported to achieve better efficiency and shorter analysis times, probably because of the competition between the solute and the similarly charged IPR [139–142]. Actually this strategy mimics the collaborative behavior of the cation and the anion when ILs are used as IPRs.

7.6   IPR COUNTER IONS

In IPC, the potential difference between the stationary phase and the bulk eluent develops because of the different adsorbophilicities of the anion and cation of the IPR. The potential determining ion is the ionic species of the IPR that experiences the strongest tendency toward adsorption onto the stationary phase; it may be either the anion or the cation. The potential determining ion is the actual ion-pairing reagent.

For sodium dodecanesulfonate, a classical IPR for positively charged analytes, the potential determining ion is dodecanesulfonate, while the counter ion is the sodium cation. The electrified interphase will carry a barely negative charge since the driving force for sodium cation adsorption is much lower than that for dodecane-sulfonate. Conversely, for tetrabutylammonium chloride, a typical IPR for negatively charged analytes, the potential determining ion is tetrabutylammonium, while the counter ion is chloride anion. The electrified interphase will bear a barely positive charge. The nature of the counter ion was demonstrated to be very important to regulate analyte retention since adsorbophilic counter ions reduce the net surface charge that results from potential determining ion adsorption.

The adsorbophilicity of a classical IPR inorganic counter ion is usually much lower than that of the organic ion that serves as the potential determining ion. For example, we can assume that the adsorbophilicity of Na^+ is negligible compared to that of dodecanesulfonate. If the IPR is a chaotropic salt, e.g. sodium perchlorate, the adsorbophilicity of the anion is stronger than that of the cation but the latter cannot be completed neglected [8]. This should be taken into account, including the surface excess of Na^+ when calculating surface charge density. Something similar occurs with small volatile IPRs used for IPC-MS as ammonium acetates or formates. The mutual contribution of the cationic and anionic components of the IPR salt was evident when ionic liquids were used as IPRs [122,124,128] as discussed above. Hence, when an IPR is used, we should distinguish between the potential determining ion H and its C counter ion.

The presence of adsorbophilic counter ions on the surface reduces the surface potential difference since C charges partially counterbalance the opposite charges of the adsorbed H ions. The surface potential has to be calculated according to the following expression:

$$\psi^0 = \frac{2RT}{F} \ln \left\{ \frac{[LH]_{Z_H} F + [LC]_{Z_C} F}{\left(8\varepsilon_0\varepsilon_r RT \sum_i c_{0i}\right)} + \left[\frac{([LH]_{Z_H} F + [LC]_{Z_C} F)^2}{8\varepsilon_0\varepsilon_r RT \sum_i c_{0i}} + 1 \right]^{\frac{1}{2}} \right\} \quad (7.1)$$

Equation 7.1 parallels Equation 3.5 (see Chapter 3, also for symbols meaning) that holds when $[LC]$, that is the surface concentration of the counterions C, is negligible compared to $[LH]$, as in the case of most traditional IPRs.

Since the amount of Na^+ paired to the adsorbed ClO_4^- is not easily measurable, we may expect that the surface excess of ClO_4^- is a fraction of the total adsorbed $NaClO_4$ that can be determined via the adsorption isotherm.

The nature of the counter ion is important because it establishes the exact value of the potential difference between the stationary phase and the bulk eluent, and also, if it exhibits suitable absorption characteristics, indirect photometric detection of UV-inactive analytes is feasible. Moreover, the choice of the counter ion of the potential determining ion allows a tailor-made separation of the analytes since adsorbophilic counter ions may compete with the analyte for interaction with the potential determining ion, thereby decreasing analyte retention. Different counter ions may alter the elution sequences of a series of analytes with potential advantage for resolution and identification purposes [143].

7.7 CONCLUSIONS

We have demonstrated that mobile phase amendment with a range of different IPRs represents a diachronic scientific consideration. It can be speculated that any charged species added to a mobile phase, may play the IPR role; they interact with the stationary phase, establishing electrostatic potentials according to their specific adsorption isotherms. It follows that charged analyte retention is altered via both electrostatic interactions with the stationary phase and pairing equilibria in bulk eluent. This indicates how broad in scope and versatile IPC is.

REFERENCES

1. Knox, J.H. and Laird, G.R. Soap chromatography: a new technique for separation of ionisable materials: dyestuff intermediates. *J. Chromatogr.* 1976, 122, 17–34.
2. Bidlingmeyer, B.A. et al. Retention mechanism for reversed-phase ion-pair liquid chromatography. *J. Chromatogr.* 1979, 186, 419–434.
3. Zappoli, S., Morselli, L., and Osti, F. Application of ion interaction chromatography to the determination of metal ions in natural water samples. *J. Chromatogr. A.* 1996, 721, 269–277.
4. Cecchi, T., Pucciarelli, F., and Passamonti, P. Extended thermodynamic approach to ion-interaction chromatography. *Anal. Chem.* 2001, 73, 2632–2639.
5. Adoubel, A. et al. Separation of underivatized small peptides on a porous graphitic carbon column by ion-pair chromatography and evaporative light scattering detection. *J. Liq. Chromatogr. Rel. Technol.* 2000, 23, 2433–2446.
6. Chaimbault, P. et al. Ion-pair chromatography on a porous graphitic carbon stationary phase for analysis of 20 underivatized protein amino acids. *J. Chromatogr. A.* 2000, 870, 245–254.
7. Shibue, M., Mant, C.T., and Hodges, R.S. Effect of anionic ion-pairing reagent concentration (1–60 mM) on reversed-phase liquid chromatography elution behaviour of peptides. *J. Chromatogr. A.* 2005, 1080, 58–67.
8. Bartha, A. and Stahlberg, J. Electrostatic retention model of reversed-phase ion-pair chromatography. *J. Chromatogr. A.* 1994, 668, 255–284.
9. Wybraniec, S. Effect of tetraalkylammonium salts on retention of betacyanins and decarboxylated betacyanins in ion-pair reversed-phase high-performance liquid chromatography. *J. Chromatogr. A.* 2006, 1127, 70–75.
10. Shibue, M., Mant, C.T., and Hodges, R.S. Effect of anionic ion-pairing reagent hydrophobicity on selectivity of peptide separations by reversed-phase liquid chromatography. *J. Chromatogr. A.* 2005, 1080, 68–75.
11. Knox, J.H. and Hartwick, R.A. Mechanism of ion-pair liquid chromatography of amines, neutrals, zwitterions and acids using anionic hetaerons. *J. Chromatogr.* 1981, 204, 3–21.
12. Lu, B., Jonsson, P., and Blomberg, S. Reversed phase ion-pair high performance liquid chromatographic gradient separation of related impurities in 2,4-disulfonic acid benzaldehyde di-sodium salt. *J. Chromatogr. A.* 2006, 1119, 270–276.
13. Pelletier, S. and Lucy, C.A. Fast and high-resolution ion chromatography at high pH on short columns packed with 1.8 μm surfactant-coated silica reverse-phase particles. *J. Chromatogr. A.* 2006, 1125, 189–194.
14. Sivaraman, N. et al. Separation of lanthanides using ion-interaction chromatography with HDEHP coated columns. *J. Radioanal. Nucl. Chem.* 2002, 252, 491–495.
15. LoBrutto, R. et al. Effect of the eluent pH and acidic modifiers in high-performance liquid chromatography retention of basic analytes. *J. Chromatogr. A.* 2001, 913, 173–187.

16. Mant, C.T. and Hodges. R.S. Context-dependent effects on the hydrophilicity and hydrophobicity of side-chains during reversed-phase high-performance liquid chromatography: implications for prediction of peptide retention behaviour. *J. Chromatogr. A.* 2006, 1125, 211–219.

17. Megoulas N.C. and Koupparis, M.A. Enhancement of evaporative light scattering detection in high-performance liquid chromatographic determination of neomycin based on highly volatile mobile phase, high-molecular-mass ion-pairing reagents and controlled peak shape. *J. Chromatogr. A.* 2004, 1057, 125–131.

18. Li, B. et al. Chracterization of impurities in tobramycin by liquid chromatography-mass spectrometry. *J. Chromatogr. A.* 2009, 1216, 3941–3945.

19. Petritis, K. et al. Parameter optimization for the analysis of underivatized protein amino acids by liquid chromatography and ionspray tandem mass spectrometry. *J. Chromatogr. A.* 2000, 896, 253–263.

20. Petritis, K. et al. Ion-pair reversed-phase liquid chromatography–electrospray mass spectrometry for the analysis of underivatized small peptides. *J. Chromatogr. A.* 2002, 957, 173–185.

21. Petritis, K.N. et al. Ion-pair reversed-phase liquid chromatography for determination of polar underivatized amino acids using perfluorinated carboxylic acids as ion-pairing agent. *J. Chromatogr. A.* 1999, 833, 147–155.

22. Castro, R., Moyano, E., and Galceran, M.T. Ion-pair liquid chromatography–atmospheric pressure ionization mass spectrometry for the determination of quaternary ammonium herbicides. *J. Chromatogr. A.* 1999, 830, 145–154.

23. Castro, R., Moyano, E., and Galceran, M.T. On-line ion-pair solid-phase extraction–liquid chromatography–mass spectrometry for the analysis of quaternary ammonium herbicides. *J. Chromatogr. A.* 2000, 869, 441–449.

24. Petritis, K. et al. Validation of ion-interaction chromatography analysis of underivatized amino acids in commercial preparation using evaporative light scattering detection. *Chromatographia* 2004, 60, 293–298.

25. Ariffin, M.M. and Anderson, R.A. LC/MS/MS analysis of quaternary ammonium drugs and herbicides in whole blood. *J. Chromatogr. B.* 2006, 842, 91–97.

26. Aramendia, M.A. et al. Determination of diquat and paraquat in olive oil by ion-pair liquid chromatography–electrospray ionization mass spectrometry. *Food Chem.* 2006, 97, 181–188.

27. Wybraniec, S. and Mizrahi, Y. Influence of perfluorinated carboxylic acids on ion-pair reversed-phase high-performance liquid chromatographic separation of betacyanins and 17-decarboxy-betacyanins. *J. Chromatogr. A.* 2004, 1029, 97–101.

28. Häkkinen, M.R. et al. Analysis of underivatized polyamines by reversed phase liquid chromatography with electrospray tandem mass spectrometry. *J. Pharm. Biomed. Anal.* 2007, 45, 625–634.

29. Gao, S. et al. Evaluation of volatile ion-pair reagents for the liquid chromatography–mass spectrometry analysis of polar compounds and its application to the determination of methadone in human plasma. *Pharm. Biomed. Anal.* 2006, 40, 679–688.

30. Jerz, G. et al. Separation of betalains from barries of *Phytolacca americana* by ion-pair high-speed counter-current chromatography. *J. Chromatogr. A.* 2008, 1190, 63–73.

31. Chen, X. et al. Simultaneous determination of amodiaquine and its active metabolite in human blood by ion-pair liquid chromatography-tandem mass spectrometry. *J. Chromatogr. B.* 2007, 860, 18–25.

32. Popa, T.V., Mant, C.T., and Hodges, R.S. Capillary electrophoresis of cationic random coil peptide standards: effect of anionic ion-pairing reagents and comparison with reversed-phase chromatography. *Electrophoresis* 2004, 25, 1219–1229.

33. Chen, Y. et al. Optimum concentration of trifluoroacetic acid for reversed-phase liquid chromatography of peptides revisited. *J. Chromatogr. A.* 2004, 1043, 9–18.

34. Kapolna, E. et al. Selenium speciation studies in Se-enriched chives (*Allium schoeno-prasum*) by HPLC-ICP–MS. *Food Chem.* 2007, 101, 1398–1406.
35. Cankur, O. et al. Selenium speciation in dill (*Anethum graveolens* L.) by ion-pairing reversed phase and cation exchange HPLC with ICP-MS detection. *Talanta* 2006, 70, 784–790.
36. Montes-Bayon, M. et al. Evaluation of different sample extraction strategies for selenium determination in selenium-enriched plants (*Allium sativum* and *Brassica juncea*) and Se speciation by HPLC-ICP-MS. *Talanta* 2006, 68, 1287–1293.
37. Fuh, M.R., Wu, T.Y., and Lin, T.Y. Determination of amphetamine and methamphetamine in urine by solid phase extraction and ion-pair liquid chromatography–electrospray tandem mass spectrometry. *Talanta* 2006, 68, 987–991.
38. Loffler, D. and Ternes, T.A. Analytical method for the determination of aminoglycoside gentamicin in hospital wastewater via liquid chromatography–electrospray tandem mass spectrometry. *J. Chromatogr. A.* 2003, 1000, 583–588.
39. Sun, D. et al. Measurement of stable isotopic enrichment and concentration of long-chain fatty acyl-carnitines in tissue by HPLC-MS. *J. Lipid Res.* 2006, 47, 431–439.
40. Gergely, V. et al. Selenium speciation in *Agaricus bisporus* and *Lentinula edodes* mushroom proteins using multi-dimensional chromatography coupled to inductively coupled plasma mass spectrometry. *J. Chromatogr. A.* 2006, 1101, 94–102.
41. Vonderheide, A.P. et al. Investigation of selenium-containing root exudates of *Brassica juncea* using HPLC-ICP-MS and ESI-qTOF-MS. *Analyst* 2006, 131, 33–40.
42. Chaimbault, P. et al. Determination of 20 underivatized proteinic amino acids by ion-pairing chromatography and pneumatically assisted electrospray mass spectrometry. *J. Chromatogr. A.* 1999, 855, 191–202.
43. Piraud, M. et al. Ion-pairing reversed-phase liquid chromatography/electrospray ionization mass spectrometric analysis of 76 underivatized amino acids of biological interest: a new tool for the diagnosis of inherited disorders of amino acid metabolism. *Rapid Commun. Mass Spectrom.* 2005, 19, 1587–1602.
44. Takino, M., Daishima, S., and Yamaguchi, K. Determination of diquat and paraquat in water by liquid chromatography/electrospray mass spectrometry using volatile ion-pairing reagents. *Anal Sci.* 2000, 16, 707–711.
45. Gustavsson, S.A. et al. Studies of signal suppression in liquid chromatography–electrospray ionization mass spectrometry using volatile ion-pairing reagents. *J. Chromatogr. A.* 2001, 937, 41–47.
46. Fuh, M.R. et al. Determination of free-form amphetamine in rat brain by ion-pair liquid chromatography–electrospray mass spectrometry with in vivo microdialysis. *J. Chromatogr. A.* 2004, 1031, 197–201.
47. Keever, J., Voyksner, R.D., and Tyczkowska, K.L. Quantitative determination of ceftiofur in milk by liquid chromatography–electrospray mass spectrometry. *J. Chromatogr. A.* 1998, 794, 57–62.
48. Kwon, J. and Moini, M. Analysis of underivatized amino acid mixtures using high performance liquid chromatography/dual oscillating nebulizer atmospheric pressure microwave induced plasma ionization mass spectrometry. *J. Am. Soc. Mass Spectrom.* 2001, 12, 117–122.
49. Toll, H. et al. Separation, detection, and identification of peptides by ion-pair reversed-phase high-performance liquid chromatography-electrospray ionization mass spectrometry at high and low pH. *J. Chromatogr. A.* 2005, 1079, 274–286.
50. Sarri, A.K., Megoulas, N.C., and Koupparis, M.A. Development of a novel liquid chromatography evaporative light scattering detection method for bacitracins and applications to quality control of pharmaceuticals. *Anal. Chim. Acta* 2006, 573, 250–257.

51. De Miguel, I., Puech-Costes, E., and Samain, D. Use of mixed perfluorinated ion-pairing agents as solvents in ion-pair high-performance liquid chromatography for the preparative purification of aminoglycoside antibiotics. *J. Chromatogr.* 1987, 407, 109–119.

52. Häkkinen, M.R. et al. Quantitative determination of underivatized polyamines by using isotope dilution RP-LC-ESI-MS/MS. *J. Pharm. Biomed. Anal.* 2008, 48, 414–421.

53. Luo, H. et al. Application of silica-based hyper-cross linked sulfonate-modified reversed stationary phases for separating highly hydrophilic basic compounds. *J. Chromatogr. A.* 2008, 1202, 8–18.

54. Gavin, P.F. and Olsen, B.A. A quality by design approach to impurity method development for atomoxetine hydrochloride (LY139603). *J. Pharm. Biomed. Anal.* 2008, 46, 431–441.

55. Holcapek, M., Jandera, P., and Zderadicka, P. High performance liquid chromatography–mass spectrometric analysis of sulphonated dyes and intermediates. *J. Chromatogr. A.* 2001, 926, 175–186.

56. Rafols, C. and Barcelo, D. Determination of mono- and disulphonated azo dyes by liquid chromatography–atmospheric pressure ionization mass spectrometry. *J. Chromatogr. A.* 1997, 777, 177–192.

57. Baiocchi, C. et al. Characterization of methyl orange and its photocatalytic degradation products by HPLC/UV–VIS diode array and atmospheric pressure ionization quadrupole ion trap mass spectrometry. *Int. J. Mass Spectrom.* 2002, 214, 247–256.

58. Holcapek, M., Jandera, P., and Prikryl, J. Analysis of sulphonated dyes and intermediates by electrospray mass spectrometry. *Dyes Pigments* 1999, 43, 127–137.

59. Lemr, K. et al. Analysis of metal complex azo dyes by high-performance liquid chromatography/electrospray ionization mass spectrometry and multistage mass spectrometry. *Rapid Commun. Mass Spectrom.* 2000, 14, 1881–1888.

60. Bharathi, V.D. et al. LC-MS-MS assay for simultaneous quantification of fexofenadine and pseudoephedrine in human plasma. *Chromatographia* 2008, 67, 461–466.

61. Liu, Y.Q. et al. Quantitative determination of erythromycylamine in human plasma by liquid chromatography-mass spectrometry and its application in a bioequivalence study of dirithromycin. *J. Chromatogr. A.* 2008, 864, 1–8.

62. Saber, A.L. et al. Liquid chromatographic and potentiometric methods for deteminations of clopidogrel *J. Food Drug Anal.* 2008, 16, 11–18.

63. Vas, G. et al. Study of transaldolase deficiency in urine samples by capillary LC-MS/MS. *J. Mass Spectrom.* 2006, 41, 463–469.

64. Ansorgov, D., Holcapek, M., and Jandera, P. Ion-pairing high-performance liquid chromatography–mass spectrometry of impurities and reduction products of sulphonated azo dyes. *J. Sep. Sci.* 2003, 26, 1017–1027.

65. Coulier, L. et al. Simultaneous quantitative analysis of metabolites using ion-pair liquid chromatography–electrospray ionization mass spectrometry. *Anal. Chem.* 2006, 78, 6573–6582.

66. Loos, R. and Barcelo, D. Determination of haloacetic acids in aqueous environments by solid-phase extraction followed by ion-pair liquid chromatography–electrospray ionization mass spectrometric detection. *J. Chromatogr. A.* 2001, 938, 45–55.

67. Sasaki, T., Iida, T., and Nambara, T. High-performance ion-pair chromatographic behaviour of conjugated bile acids with di-n-butylamine acetate. *J. Chromatogr. A.* 2000, 888, 93–102.

68. Fer, M. et al. Determination of polyunsatured fatty acid monoepoxides by high performance liquid chromatography–mass spectrometry. *J. Chromatogr. A.* 2006, 1115, 1–7.

69. Holcapek, M. et al. Effects of ion-pairing reagents on the electrospray signal suppression of sulphonated dyes and intermediates. *J. Mass Spectrom.* 2004, 39, 43–50.

70. Storm, T., Reemtsma, T., and Jekel, M. Use of volatile amines as ionpairing agents for the HPLC–tandem mass spectrometric determination of aromatic sulphonates in industrial wastewater. *J. Chromatogr. A.* 1999, 854, 175–195.

71. Reemtsma, T. Analysis of sulphophthalimide and some of its derivatives by liquid chromatography–electrospray ionization tandem mass spectrometry. *J. Chromatogr. A.* 2001, 919, 289–297.

72. Tak, V. et al. Application of Doehlert design in optimizing the determination of degraded products of nerve agents by ion-pair liquid chromatography electrospray ionization tandem mass spectrometry, *J. Chromatogr. A.* 2007, 1161, 198–206.

73. Reemtsma, T. The use of liquid chromatography atmospheric pressure ionization mass spectrometry in water analysis. *Trends Anal. Chem.* 2001, 20, 500–517.

74. Alonso, M.C. and Barcelo, D. Tracing polar benzene- and naphthalenesulfonates in untreated industrial effluents and water treatment works by ion-pair chromatography–fluorescence and electrospray mass spectrometry. *Anal. Chim. Acta* 1999, 400, 211–231.

75. Schmidt, T.C, Buetehorn, U., and Steinbach, K. HPLC-MS investigations of acidic contaminants in ammunition wastes using volatile ion-pairing reagents (VIP-LC-MS). *Anal. Bioanal. Chem.* 2004, 378, 926–931.

76. Fountain, K.J., Gilar, M., and Gebler, J.C. Analysis of native and chemically modified oligonucleotides by tandem ion-pair reversed-phase high-performance liquid chromatography/electrospray ionization mass spectrometry. *Rapid Commun. Mass Spectrom.* 2003, 17, 646–653.

77. Shen, X. et al. Investigation of copper azamacrocyclic complexes by high-performance liquid chromatography. *Biomed. Chromatogr.* 2006, 20, 37–47.

78. Kuberan, B. et al. Analysis of heparan sulfate oligosaccharides with ion-pair reverse phase capillary high performance liquid chromatography–microelectrospray ionization time-of-flight mass spectrometry. *J. Am. Chem. Soc.* 2002, 124, 8707–8718.

79. Vanerkova, D. et al. Analysis of electrochemical degradation products of sulphonated azo dyes using high-performance liquid chromatography/tandem mass spectrometry. *Rapid Commun. Mass Spectrom.* 2006, 20, 2807–2815.

80. Cordell, R.L. et al. Quantitative profiling of nucleotides and related phosphate-containing metabolites in cultured mammalian cells by liquid chromatography tandem electrospray mass spectrometry. *J. Chromatogr. B.* 2008, 871, 115–124.

81. Lue, B.M., Guo, Z., and Xu, X. High-performance liquid chromatography analysis methods developed for quantifying enzymatic esterification of flavonoids in ionic liquids. *J. Chromatogr. A.* 2008, 1198, 107–114.

82. Narayanaswamy, M. et al. An assay combining high-performance liquid chromatography and mass spectrometry to measure DNA interstrand cross-linking efficiency in oligonucleotides of varying sequences. *Anal. Biochem.* 2008, 374, 173–181.

83. Pruvost, A. et al. Specificity enhancement with LC-positive ESI-MS/MS for the measurement of nucleotides. *J. Mass Spectrom.* 2008, 43, 224–233.

84. Gibson, C.R. et al. Electrospray ionization mass spectrometry coupled to reversed-phase ion-pair high-performance liquid chromatography for quantification of sodium borocaptate and application to pharmacokinetic analysis. *Anal. Chem.* 2002, 74, 2394–2399.

85. Witters, E. et al. Ion-pair liquid chromatography–electrospray mass spectrometry for the analysis of cyclic nucleotides. *J. Chromatogr. B.* 1997, 694, 55–63.

86. Ishikawa, A. and Shibata, T. Cellulosic chiral stationary phase under reversed-phase condition. *J. Liq. Chromatogr.* 1993, 16, 859–878.

87. Flieger J. Effect of chaotropic mobile phase additives on the separation of selected alkaloids in reversed-phase high-performance liquid chromatography. *J. Chromatogr. A.* 2006, 1113, 37–44.

88. Kazakevich, I.L. and Snow, N.H. Adsorption behavior of hexafluorophosphate on selected bonded phases. *J. Chromatogr. A.* 2006, 1119, 43–50.

89. Hashem, H. and Jira, T. Effect of chaotropic mobile phase additives on retention behaviour of β-blockers on various reversed-phase high-performance liquid chromatography columns. *J. Chromatogr. A.* 2006, 1133, 69–75.

90. Basci, N.E. et al. Optimization of mobile phase in the separation of β-blockers by HPLC. *J. Pharm. Biomed. Anal.* 1998, 18, 745–750.

91. LoBrutto, R. et al. Effect of the eluent pH and acidic modifiers in high-performance liquid chromatography retention of basic analytes. *J. Chromatogr. A.* 2001, 913, 173–187.

92. Loeser, E. and Drumm, P. Investigation of anion retention and cation exclusion effects for several C18 stationary phases. *Anal. Chem.* 2007, 79, 5382–5391.

93. Harrison, C.R, Sader, J.A, and Lucy, C.A. Sulfonium and phosphonium, new ion-pairing agents with unique selectivity toward polarizable anions. *J. Chromatogr. A.* 2006, 1113, 123–129.

94. Gritti, F. and Guiochon, G. Effect of the pH, concentration and nature of the buffer on the adsorption mechanism of an ionic compound in reversed-phase liquid chromatography II: analytical and overloaded band profiles on Symmetry-C18 and Xterra-C18. *J. Chromatogr. A.* 2004, 1041, 63–75.

95. Jones, A., LoBrutto, R., and Kazakevich, Y.V. Effect of the counter-anion type and concentration on the liquid chromatography retention of β-blockers. *J. Chromatogr. A.* 2002, 964, 179–187.

96. Courderot, C.M. et al. Chiral discrimination of dansyl-amino-acid enantiomers on teicoplanin phase: sucrose–perchlorate anion dependence. *Anal. Chim. Acta* 2002, 457, 149–155.

97. Pilorz, K. and Choma, I. Isocratic reversed-phase high-performance liquid chromatographic separation of tetracyclines and flumequine controlled by a chaotropic effect. *J. Chromatogr. A.* 2004, 1031, 303–305.

98. Crespi, C.L., Chang, T.K., and Waxman, D.J. CYP2D6-dependent bufuralol 1′-hydroxylation assayed by reverse-phase ion-pair high-performance liquid chromatography with fluorescence detection. *Methods Mol. Biol.* 2006, 320, 121–125.

99. Bogusz, M. *Journal of Chromatography Library.* Elsevier: Amsterdam, 1995, Chap. 5.

100. Hashem, H. and Jira, T. Retention behaviour of β-blockers in HPLC using a monolithic column. *J. Sep. Sci.* 2006, 29, 986–994.

101. Roberts, J.M. et al. Influence of the Hofmeister series on retention of amines in reversed-phase liquid chromatography. *Anal. Chem.* 2002, 74, 4927–4932.

102. Gritti, F. and Guiochon, G. Effect of the ionic strength of salts on retention and overloading behaviour of ionizable compounds in reversed-phase liquid chromatography I: XTerra-C18. *J. Chromatogr. A.* 2004, 1033, 43–55.

103. Gritti, F. and Guiochon, G. Role of the buffer in retention and adsorption mechanism of ionic species in reversed-phase liquid chromatography I: analytical and overloaded band profiles on Kromasil-C18. *J. Chromatogr. A.* 2004, 1038, 53–66.

104. Gritti, F. and Guiochon, G. Effect of the ionic strength of salts on retention and overloading behaviour of ionizable compounds in reversed-phase liquid chromatography II: Symmetry-C18. *J. Chromatogr. A.* 2004, 1033, 57–69.

105. Shibue, M., Mant, C.T., and Hodges, R.S. The perchlorate anion is more effective than trifluoroacetate anion as an ion-pairing reagent for reversed-phase chromatography of peptides. *J. Chromatogr. A.* 2005, 1080, 49–57.

106. Kazakevich, Y.V., LoBrutto, R., and Vivilecchia, R. Reversed-phase high-performance liquid chromatography behavior of chaotropic counter anions. *J. Chromatogr. A.* 2005, 1064, 9–18.

107. Cecchi, T. and Passamonti, Retention Mechanism of ion-pair chromatography with chaotropic additives. *J. Chromatogr. A.* 2009, 1216, 789–797. and Erratom in *J. Chromatogr. A* 216, 5164.

108. Pan, L. et al. Influence of inorganic mobile phase additives on the retention, efficiency and peak symmetry of protonated basic compounds in reversed-phase liquid chromatography. *J. Chromatogr. A.* 2004, 1049, 63–73.

109. Rodriguez-Nogales, J.M., Garcia, M.C., and Marina, M.L. Development of a perfusion reversed-phase high performance liquid chromatography method for the characterisation of maize products using multivariate analysis. *J. Chromatogr. A.* 2006, 1104, 91–99.

110. Castro-Rubio, A. et al. Determination of soybean proteins in soybean–wheat and soybean–rice commercial products by perfusion reversed phase high-performance liquid chromatography. *Food Chem.* 2007, 100, 948–955.

111. Fanciulli, G. et al. Quantification of gluten exorphin A5 in cerebrospinal fluid by liquid chromatography–mass spectrometry. *J. Chromatogr. B.* 2006, 833, 204–209.

112. Shapovalova, E.N. et al. Ion-pair chromatography of metal complexes of unithiol in the presence of quaternary phosphonium salts. *J. Anal. Chem.* 2001, 56, 160–165.

113. Holbrey, J.D. and Seddon, K.R. Ionic liquids. *Clean Prod. Processes* 1999, 1, 223–236.

114. Liu, J., Jönsson, J.A., and Jiang, G. Application of ionic liquids in analytical chemistry. *Trends Anal. Chem.* 2005, 24, 20–27.

115. Stepnowski, P. Application of chromatographic and electrophoretic methods for analysis of imidazolium and pyridinium cations as used in ionic liquids. *Int. J. Mol. Sci.* 2006, 7, 497–509.

116. Knox, J.H., Kaliszan, R., and Kennedy, G.J. Enthalpic exclusion chromatography. *Faraday Discuss. Roy. Chem. Soc.* 1980, 15, 113–125.

117. Reta, M. and Carr, P.W. Comparative study of divalent metals and amines as silanol-blocking agents in reversed-phase liquid chromatography. *J. Chromatogr. A.* 1999, 855, 121–127.

118. Kaliszan, R. et al. Suppression of deleterious effects of free silanols in liquid chromatography by imidazolium tetrafluoroborate ionic liquids. *J. Chromatogr. A.* 2004, 1030, 263–271.

119. Marszałł, M.P. and Kaliszan, R. Application of ionic liquids in liquid chromatography. *Crit. Rev. Anal. Chem.* 2007, 37, 127–140.

120. Marszałł, M.P., Bączek, T., and Kaliszan, R. Evaluation of the silanol-supressing potency of ionic liquids. *J. Sep. Sci.* 2006, 29, 1138–1145.

121. Marszałł, M.P., Bączek, T., and Kaliszan, R. Reduction of silanophilic interactions in liquid chromatography with the use of ionic liquids. *Anal. Chim. Acta* 2005, 547, 172–178.

122. Ruiz-Angel, M.J., Carda-Broch, S., and Berthod A. Ionic liquids versus triethylamine as mobile phase additives in the analysis of β-blockers. *J. Chromatogr. A.* 2006, 1119, 202–208.

123. Waichigo, M.M., Riechel, T.L., and Danielson, N.D. Ethylammonium acetate as a mobile phase modifier for reversed phase liquid chromatography. *Chromatographia* 2005, 61, 17–23.

124. Berthod, A., Ruiz-Angel, M.J., and Huguet, S. Nonmolecular solvents in separation methods: dual nature of room temperature ionic liquids. *Anal. Chem.* 2005, 77, 4071–4080.

125. Gritti, F. and Guiochon, G. Retention of ionizable compounds in reversed-phase liquid chromatography: effect of ionic strength of the mobile phase and the nature of salts used on overloading behavior. *Anal. Chem.* 2004, 76, 4779–4789.

126. Waichigo, M.M. and Danielson, N.D. Ethylammonium formate as an organic solvent replacement for ion-pair reversed-phase liquid chromatography. *J. Chromatogr. Sci.* 2006, 44, 607–614.

127. He, L. et al. Effect of 1-alkyl-3-methylimidazolium-based ionic liquids as the eluent on the separation of ephedrines by liquid chromatography. *J. Chromatogr. A.* 2003, 1007, 39–45.

128. Tang, F. et al. Determination of octopamine, synephrine and tyramine in Citrus herbs by ionic liquid improved 'green' chromatography. *J. Chromatogr. A.* 2006, 1125, 182–188.

129. Jin, C.H., Polyakova, Y., and Kyung, H.R. Effect of concentration of ionic liquids on resolution of nucleotides in reversed-phase liquid chromatography. *Bull. Kor. Chem. Soc.* 2007 28, 601–606.

130. Yoo, C.G., Han, Q., and Mun, S. Reversal of elution sequence and selectivity resulting from the use of an ionic liquid as a mobile phase modifier. *J. Liq. Chromatogr. Rel. Technol.* 2008, 31, 1104–1122.

131. Xiao, X. et al. S. Ionic liquids as additives in high performance liquid chromatography: analysis of amines and the interaction mechanism of ionic liquids. *Anal. Chim. Acta* 2004, 519, 207–211.

132. Polyakova, Y., Koo, Y.M., and Row, K.H. Application of ionic liquids as mobile phase modifier in HPLC. *Biotechnol. Bioprocess. Eng.* 2006, 11, 1–6.

133. Woollard, D.C. and Indyk-Harvey, E. Rapid determination of thiamine, riboflavin, pyridoxine, and niacinamide in infant formulas by liquid chromatography. *J. AOAC Int.* 2002, 85, 945–951.

134. Huang, Z. et al. Determination of cyclamate in foods by high performance liquid chromatography–electrospray ionization mass spectrometry. *Anal. Chim. Acta* 2006, 555, 233–237.

135. Bruzzoniti, M.C., Mentasti, E., and Sarzanini, C. Divalent pairing ion for ion-interaction chromatography of sulfonates and carboxylates. *J. Chromatogr. A.* 1997, 770, 51–57.

136. Tonelli, D., Zappoli, S., and Ballarin, B. Dye-coated stationary-phases: a retention model for anions in ion interaction chromatography. *Chromatographia* 1998, 48, 190–196.

137. Papoušková, B. et al. Mass spectrometric study of selected precursors and degradation products of chemical warfare agents. *J. Mass Spectrom.* 2007, 42, 1550–1561.

138. Cecchi, T., Pucciarelli F., and Passamonti P. Ion interaction chromatography of zwitterions: the fractional charge approach to model the influence of mobile phase concentration of the ion-interaction reagent. *Analyst* 2004, 129, 1037–1042.

139. Yalcın, G. and Yuktas, F.N. An efficient separation and method development for the quantifying of two basic impurities of nicergoline by reversed-phase high performance liquid chromatography using ion-pairing counter ions. *J. Pharm. Biomed. Anal.* 2006, 42, 434–440.

140. Fletouris, D.J. et al. Highly sensitive ion-pair liquid chromatographic determination of albendazole marker residue in animal tissues. *J. Agr. Food Chem.* 2005, 53, 893–898.

141. Fletouris, D.J., Papapanagiotou, E.P. A new liquid chromatographic method for routine determination of oxytetracycline marker residue in the edible tissues of farm animals. *Anal. Bioanal. Chem.* 2008, 391, 1189–1198.

142. Amin, A.S., Shahat, M.F.E1., Eden, R.E., Meshref, M.A. *Anal. Lett.* Comparison of ion-pairing and reversed phase chromatography in determination of sulfmethoxazole and trimethoprim. 2008, 41(10), 1878–1894.

143. Gennaro, M.C. *Advances in Chromatography.* Reversed-Phase Ion-Pair and Ion-Interaction Chromatography, 1995, 35(7), 343–381.

8 Organic Modifiers

8.1 ORGANIC MODIFIER CONCENTRATION IN ELUENT

In reversed phase chromatography, the dependence of solute retention on the organic modifier concentration (φ) in the mobile phase is sanctioned by the following relationship [1]:

$$\ln k_\varphi = \ln k_w - S\varphi \qquad (8.1)$$

where k_w is the retention factor when $\varphi = 0\%$ and S is a constant for a given solute–solvent combination and is related to the free energy of transferring the solute from the pure aqueous to the pure organic phase. It is clear from Equation 8.1 that the free energy of adsorption is a linear function of organic modifier concentration.

Clearly, both analyte retention and the ion-pairing reagent (IPR) adsorption isotherm are weakened in organic modifier-rich eluents. Since the goal of the IPC strategy (usually performed under reversed phase conditions) is to improve analyte retention, it follows that IPC encompasses the mobile phase region characterized by water-rich solutions. In spite of this, the presence of at least small amounts of organic modifier in the mobile phase is crucial to enhance the wetting of the reversed phase chromatographic packing. This avoids the phase collapse of the bonded ligands due to hydrophobic staking of the alkyl chains under 100% water eluent. Usually the correct percentage of organic modifier in the eluent results from a trade-off between the eluent strength (that should be as low as possible to obtain adequate retention) and the analyte solubility in the mobile phase. However in recent applications, it has not been unusual to adjust the separation selectivity via fine tuning of this parameter.

The addition of an organic modifier to the aqueous eluent modifies its viscosity and lowers its surface tension. As explained in Chapter 3 (Section 3.1.2), by increasing the organic modifier eluent concentration, a quantitatively predictable [2] decrease of the retention factor of the analyte is obtained, as proven by considerable experimental evidence [3–9]. The retention decrease clearly depends on the analyte nature [4].

It is instructive to observe in Figure 8.1 a family of model IPR adsorption isotherms at different mobile phase concentrations of the organic modifier according to Equation 3.7. The stationary phase coverage by the IPR decreases with increasing organic modifier in the eluent. Figure 8.2 illustrates the theoretical dependence of the retention factor of a model analyte oppositely charged to the IPR, as a function of the mobile phase concentration of the IPR, at increasing eluent percentages of the organic modifier.

Section 2.5.3 in Chapter 2 elucidated the role of the organic modifier related to ion-pairing equilibrium. The increase of the volume fraction of the organic modifier

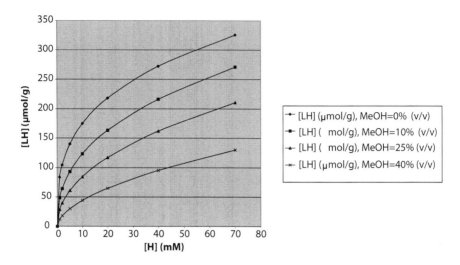

FIGURE 8.1 Family of model IPR adsorption isotherms at different mobile phase concentrations of organic modifier.

in the eluent affects electrostatic and hydrophobic interactions in conflicting ways. The former are reinforced by the decreased dielectric constant while the latter are obviously weakened since the percentage of water is decreased. If electrostatic interactions were the only driving forces in the ion-pairing process, the equilibrium constant for the duplex formation would increase with increasing organic modifier in the eluent; the decreased dielectric constant of the medium would favor the formation of a neutral species. However, when lipophilic ions (such as most classical IPRs and

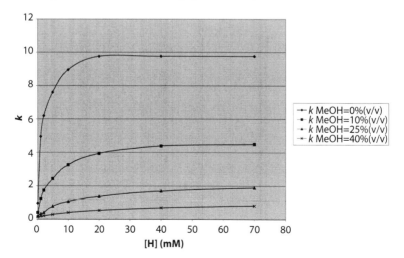

FIGURE 8.2 Theoretical dependence of retention factor of a model analyte oppositely charged to the IPR, as a function of the mobile phase concentration of the IPR, at increasing eluent percentages of organic modifier.

typical IPC analytes) are involved in the pairing process, hydrophobic interactions may lead to water-enforced ion-pairing [10].

The experimental evidence runs counter to the electrostatic description of the pairing process; the capability of water to force poorly hydrated ions together, so as to decrease their modification of the water structure, decreases with increasing percentages of the organic modifier, and amplifies with the sizes of the ions [10,11]. The stacking free energy that characterizes the interactions of hydrophobic moieties of the cation and the anion can be considered a kind of adsorption free energy that usually becomes less negative linearly with increasing concentration of the organic modifier (thereby decreasing the spontaneity of the adsorption process) [2].

Thomlinson [12] was the first chromatographer to note that the classical electrostatic ion-pair concept did not hold for bulky lipophilic IPRs; he also emphasized that in the region between the mobile and stationary phases, the dielectric constant of the medium is far lower than that of the aqueous phase. It is now clear that water-enforced pairing effects [13] include hydrophobic attraction between hydrophobic moieties of the pairing ions and dielectric saturation; actually the water-enforced pairing effects were demonstrated to be more important than electrostatic attraction even in a water–methanol system [14].

In Chapter 3 (Section 3.1.2) the theoretical predictions of a weakening of hydrophobic ion-pairing with increasing organic modifier concentration in the eluent was quantitatively confirmed via an extended thermodynamic retention model of IPC [2] that was also able to treat the simultaneous influence of the organic modifier and IPR concentrations in the eluent from a bivariate view. The retention equation correctly predicts that retention diminishes with increasing organic modifier concentration and increases with increasing IPR concentration, as expected. Interestingly, the latter increase is lower at high organic modifier eluent concentration since the organic modifier reduces analyte retention both directly (it decreases solute free energy of adsorption) and indirectly (it lessens IIR free energy of adsorption on the stationary phase; the electrostatic potential is lower, leading to a drop in retention). For inorganic chaotropic IPRs that lack hydrophobic moieties, the role of the organic modifier may be different. In the absence of an IPR, the bivariate retention equation of the extended thermodynamic approach by Cecchi and co-workers reduces to Equation 8.1 that describes the influence of the organic modifier on RP-HPLC retention.

8.2 NATURE OF ORGANIC MODIFIERS

The polarity of an organic modifier is inversely proportional to its impact on retention attenuation; different selectivities can be obtained using diverse organic modifiers [15]. The reduction of the capacity factors with increasing organic modifier concentration in the eluent was weaker when methanol was used compared to acetonitrile, and this was attributed to its lower polarity [5]. Even if methanol or acetonitrile are the most common organic modifiers, an unusual solvent, tetramethylene oxide [16], was recently tested in the IPC of sulfides and aromatic sulfonated compounds and proved to play an important role in adjusting retention.

Earlier attempts to take advantage of ionic liquids (ILs) in HPLC focused on their use to replace organic modifiers in the eluent, since their non-molecular nature could lead to original selectivities. Many experimental designs used alkylammonium ILs since their viscosity is lower that those of many other ILs even if it is at least one order of magnitude higher than those of common organic modifiers. Their very high viscosity prevents full exploitation of their potential also because the substitution of a classical organic modifier with ILs critically enhances the system back pressures and it impairs efficient mass transfer from the mobile phase to the stationary phase; this results in substandard peak efficiencies [17–19].

Two other reasons ILs have not attracted wider interest as putative organic modifier replacers are (1) their poorer UV transparency and (2) higher cost. Conversely, the successful use of these ionic compounds in IPC relies on the multiplicity of roles they can play simultaneously. The use of ILs as silanol suppressors (described in Chapter 5) and as IPRs (discussed in Chapter 7) demonstrates that they are versatile reagents in IPC as confirmed by their ability to control retention of ionized sample without including the organic modifier in the eluent. For this reason, they were proposed as environmentally friendly mobile phases [20].

The chromatograms obtained with ILs in many studies approached those achieved with methanol or acetonitrile mobile phases on the same columns, even if the retention order of the analytes was not always the same since ILs also act as IPRs. The retention behavior of neutral analytes is instructive. If the electrical charge of an analyte is zero, no ionic interactions between the analyte and the IPR are predictable. This led many model makers [21] to take for granted the constancy of retention of a neutral analyte as a function of IPR concentration. Many experimental results contradicted this hypothesis [22–24]. Only the extended thermodynamic approach by Cecchi and co-workers quantitatively modeled the relationship of neutral analyte retention and IPR concentration in the eluent [23,24]. Actually this dependence is weak and in some limiting cases may be so feeble as not to be detected. Waichigo et al. demonstrated that for acetophenone, log k linearly decreased with increasing volume fraction of ILs in the mobile phase [18] because the ionic additive impaired hydrophobic interactions between the analyte and stationary phase, thereby confirming this dependence also for unusual pairing ions such as ILs and giving credibility to the extended thermodynamic retention model of Cecchi and co-workers [25].

8.3 GRADIENT ELUTION

IPC has been traditionally performed in isocratic elution mode because the amount of the IPR adsorbed on the stationary phase strongly decreases with increasing organic modifier in the eluent. The adsorption equilibrium constant of the IPR sanctions the ratio of its equilibrium activities. in the stationary and mobile phases. For a given temperature, this constant depends on the surface chemistry of the stationary phase and on mobile phase composition. Classical IPRs are long chain sulfonates and ammonium salts. Thus, a long time may be needed to equilibrate a column with an eluent of constant composition; many column void volumes must be eluted before reproducible separation is obtained. This concept prevented pioneering IPC practitioners from focusing on the potential of the gradient elution mode because the IPR surface equilibrium concentration changes continuously with changing eluent composition and the steady state at which adsorption rate equals desorption rate cannot be reached during a run.

However, several recent examples of gradient separations of a wide range of diverse analytes prove that this strategy can be successful [4,26–52] because it avoids extreme pH conditions, brings about high peak capacity, and makes fast separations possible [53–57]. The organic modifier concentration gradient causes a change in the distribution equilibrium of the IPR between the mobile and stationary phases. The experimental evidence clearly indicates that the attainment of the surface equilibrium concentration of the IPR is not a strict prerequisite to obtain effective separations.

However, a critical review of the IPRs used in gradient elution shows that most are weakly adsorbophilic species so that the stationary phase concentration of the IPR can be more easily modulated by its eluent concentration, according to the adsorption equilibrium constant. Examples of gradient strategies with classical long chain IPRs are scarce.

Gradient elution also proved beneficial for positional isomer resolution [58]. A chemometrical approach was useful to optimize the elution gradient program of perfusion IPC for the characterization of biogenic amines in wines [26] and maize products [27].

Interestingly, many variants of the classical organic modifier gradient are adjuncts. For example a flow rate gradient was devised to shorten the cycle time of the organic modifier gradient run and achieve prompt re-equilibration of the initial column conditions—a strategy called dual mode gradient IPC [59]. A dual organic modifier concentration and pH gradient allowed the sequential separation of lanthanides [60]. Simultaneous gradients of eluent pH, organic modifier, and IPR eluent concentration led to separation of several key classes of polar metabolites. Very polar and acidic metabolites such as nucleotides were separated via an ion-pair mechanism while nonpolar compounds such as CoA esters underwent a reversed phase mechanism during the same chromatographic run [61]. Similarly, a complicated combined organic modifier, IPR and pH gradient allowed the baseline separations of alkaloids [62]. Interestingly, the switch between IPC and RP-HPLC provided by the gradients of the organic modifier and IPR eluent concentrations added selectivity (1) in the determination of fluorescent whitening agents in environmental waters [63]; (2) in the C18 reversed phase purification of 6-deoxysugar nucleotides, hexose, and pentose-derived sugar nucleotides [64]; and (3) in the measurement of reduced and total mercaptamine in urine [65]. However, simultaneous increases of both the organic modifier and the IPR were used to investigate the elution time shifts of 33 peptides and their corresponding phosphopeptides [66] and very complex mixed gradients were explored [67].

REFERENCES

1. Shoenmakers, P.J., Billiet, A.H., and De Galan, L. Influence of organic modifiers on the retention behaviour in reversed-phase liquid chromatography and its consequences for gradient elution. *J. Chromatogr.* 1979, 185, 179–195.
2. Cecchi, T., Pucciarelli, F., and Passamonti, P. Extended thermodynamic approach to ion-interaction chromatography: influence of organic modifier concentration. *Chromatographia* 2003, 5, 411–419.

3. Szabo, Z. et al. Analysis of nitrate ion in nettle (*Urtica dioica* L.) by ion-pair chromatographic method on a C30 stationary phase. *J. Agr. Food Chem.* 2006, 54, 4082–4086.

4. Lu, B., Jonsson, P., and Blomberg, S. Reversed phase ion-pair high performance liquid chromatographic gradient separation of related impurities in 2,4-disulfonic acid benzaldehyde di-sodium salt. *J. Chromatogr. A.* 2006, 1119, 270–276.

5. Zappoli, S., Morselli, L., and Osti, F. Application of ion interaction chromatography to the determination of metal ions in natural water samples. *J. Chromatogr. A.* 1996, 721, 269–277.

6. Shapovalova, E.N. et al. Ion pair chromatography of metal complexes of unithiol in the presence of quaternary phosphonium salts. *J. Anal. Chem.* 2001, 56, 160–165.

7. Bruzzoniti, M.C., Mentasti, E., and Sarzanini, C. Divalent pairing ion for ion-interaction chromatography of sulfonates and carboxylates. *J. Chromatogr. A.* 1997, 770, 51–57.

8. Podkrajsek, B., Grgic, I., and Tursic, J. Determination of sulfur oxides formed during the S(IV) oxidation in the presence of iron. *Chemosphere* 2002, 49, 271–277.

9. Sarzanini, C. et al. Sulfonated azoligand for metal ion determination in ion interaction chromatography. *J. Chromatogr. A.* 1999, 847, 233–244.

10. Diamond, R.M. The aqueous solution behaviour of large univalent ions: new type of ion-pairing. *J. Phys. Chem.* 1963, 67, 2513–2517.

11. Desnoyers, J.E. and Arel, M. Salting-in by quaternary ammonium salts. *Can. J. Chem.* 1965, 43, 3232–3237.

12. Thomlinson, E., Jefferies, T.M., and Riley, C.M. Ion pair high-performance liquid chromatography. *J. Chromatogr.* 1978, 159, 315–358.

13. Fritz, J.S. Factors affecting selectivity in ion chromatography. *J. Chromatogr. A.* 2005, 1085, 8–17.

14. Liu, L., Ouyang, J., and Baeyens, W.R.G. Separation of purine and pyrimidine bases by ion chromatography with direct conductivity detection. *J. Chromatogr. A.* 2008, 1193, 104–108.

15. Gavin, P.F. and Olsen, B.A. A quality by design approach to impurity method development for atomoxetine hydrochloride (LY139603). *J. Pharm. Biomed. Anal.* 2008, 46, 431–441.

16. Pang, X.Y., Sun, H.W., and Wang, Y.H. Determination of sulfides in synthesis and isomerization systems by reversed-phase ion-pair chromatography with a mobile phase containing tetramethylene oxide as organic modifier. *Chromatographia* 2003, 57, 543–547.

17. Waichigo, M.M. and Danielson, N.D. Comparison of ethylammonium formate to methanol as a mobile phase modifier for reversed phase liquid chromatography. *J. Sep. Sci.* 2006, 29, 599–606.

18. Waichigo, M.M., Riechel, T.L., and Danielson, N.D. Ethylammonium acetate as mobile phase modifier for reversed phase liquid chromatography. *Chromatographia* 2005, 61, 17–23.

19. Waichigo, M.M. et al. Alkylammonium formate ionic liquids as organic mobile phase replacements for reversed-phase liquid chromatography. *J. Liq. Chromatogr. Rel. Technol.* 2007, 30, 165–184.

20. Marszałł, M.P., Bączek, T., and Kaliszan, R. Evaluation of the silanol-suppressing potency of ionic liquids. *J. Sep. Sci.* 2006, 29, 1138–1145.

21. Bartha, A. and Stahlberg, J. Electrostatic retention model of reversed-phase ion-pair chromatography. *J. Chromatogr. A.* 1994, 668, 255–284.

22. Knox, J.H. and Hartwick, R.A. Mechanism of ion-pair liquid chromatography of amines, neutrals, zwitterions and acids using anionic hetaerons. *J. Chromatogr.* 1981, 204, 3–21.

23. Cecchi, T., Pucciarelli, F., and Passamonti, P. Ion interaction chromatography of neutral molecules. *Chromatographia* 2000, 53, 27–34.

24. Cecchi, T., Pucciarelli, F., and Passamonti, P. Ion interaction chromatography of neutral molecules: a potential approximation to obtain a simplified retention equation. *J. Liq. Chromatogr. Rel. Technol.* 2001, 24, 291–302.

25. Cecchi, T., Pucciarelli, F., and Passamonti, P. Extended thermodynamic approach to ion-interaction chromatography. *Anal. Chem.* 2001, 73, 2632–2639.

26. Hlabangana, L., Hernandez-Cassou, S., and Saurina J. Determination of biogenic amines in wines by ion-pair liquid chromatography and post-column derivatization with 1,2-naphthoquinone-4-sulphonate. *J. Chromatogr. A.* 2006, 1130, 130–136.

27. Rodriguez-Nogales, J.M., Garcia, M.C., and Marina, M.L. Development of a perfusion reversed-phase high performance liquid chromatography method for the characterisation of maize products using multivariate analysis. *J. Chromatogr. A.* 2006, 1104, 91–99.

28. Li J. Prediction of internal standards in reversed-phase liquid chromatography IV: correlation and prediction of retention in reversed-phase ion-pair chromatography based on linear solvation energy relationships. *Anal. Chim. Acta* 2004, 522, 113–126.

29. Li, J. and Rethwill P.A. Prediction of internal standards in reversed phase liquid chromatography. *Chromatographia* 2004, 60, 63–71.

30. Li, J. and Rethwill P.A. Systematic selection of internal standard with similar chemical and UV properties to drug to be quantified in serum samples. *Chromatographia* 2004, 60, 391–397.

31. Petritis, K. et al. Ion pair reversed-phase liquid chromatography–electrospray mass spectrometry for the analysis of underivatized small peptides. *J. Chromatogr. A.* 2002, 957, 173–185.

32. Petritis, K. et al. Validation of ion-interaction chromatography analysis of underivatized amino acids in commercial preparation using evaporative light scattering detection. *Chromatographia* 2004, 60, 293–298.

33. Garcia-Marco, S. et al. Gradient ion-pair chromatographic method for the determination of iron N,N′-ethylenediamine-di-(2-hydroxy-5-sulfophenylacetate) by high performance liquid chromatography-atmospheric pressure ionization electrospray mass spectrometry. *J. Chromatogr. A.* 2005, 1064, 67–74.

34. Castro-Rubio, A. et al. Determination of soybean proteins in soybean–wheat and soybean–rice commercial products by perfusion reversed phase high-performance liquid chromatography. *Food Chem.* 2007, 100, 948–955.

35. Ariffin, M.M. and Anderson, R.A. LC/MS/MS analysis of quaternary ammonium drugs and herbicides in whole blood. *J. Chromatogr. B.* 2006, 842, 91–97.

36. Lavizzari, T. et al. Improved method for determination of biogenic amines and polyamines in vegetable products by ion-pair high-performance liquid chromatography. *J. Chromatogr. A.* 2006, 1129, 67–72.

37. Aramendia, M.A. et al. Determination of diquat and paraquat in olive oil by ion-pair liquid chromatography–electrospray ionization mass spectrometry. *Food Chem.* 2006, 97, 181–188.

38. Qiu, J. et al. Online pre-reduction of Se(VI) by thiourea for selenium speciation by hydride generation. *Spectrochim. Acta B.* 2006, 61, 803–809.

39. Sarica, D.Y., Turker, A.R., and Erol, E. Online speciation and determination of Cr(III) and Cr(VI) in drinking and waste water samples by reversed-phase high performance liquid chromatography coupled with atomic absorption spectrometry. *J. Sep. Sci.* 2006, 29, 1600–1606.

40. Beverly, M. et al. Liquid chromatography electrospray ionization mass spectrometry analysis of the ocular metabolites from a short interfering RNA duplex. *J. Chromatogr. B.* 2006, 835, 62–70.

41. Fanciulli, G. et al. Quantification of gluten exorphin A5 in cerebrospinal fluid by liquid chromatography–mass spectrometry. *J. Chromatogr. B.* 2006, 833, 204–209.

42. Kawamura, K. et al. Separation of aromatic carboxylic acids using quaternary ammonium salts on reversed-phase HPLC 2: application for analysis of Loy Yang coal oxidation products. *Sep. Sci. Technol.* 2006, 41, 723–732.

43. Quintana, J.B., Rodil, R., and Reemtsma, T. Determination of phosphoric acid mono- and diesters in municipal wastewater by solid-phase extraction and ion-pair liquid chromatography–tandem mass spectrometry. *Anal. Chem.* 2006, 78, 1644–1650.

44. Brudin, S. et al. One-dimensional and two-dimensional liquid chromatography of sulphonated lignins. *J. Chromatogr. A.* 2008, 1201, 196–201.

45. Seifar, R.M. et al. Quantitative analysis of metabolites in complex biological samples using ion-pair reversed-phase liquid chromatography-isotope dilution tandem mass spectrometry. *J. Chromatogr. A.* 2008, 1187, 103–110.

46. Häkkinen, M.R. et al. Analysis of underivatized polyamines by reversed phase liquid chromatography with electrospray tandem mass spectrometry. *J. Pharm. Biomed. Anal.* 2007, 45, 625–634.

47. Mohn, T. et al. Extraction and analysis of intact glucosinolates: validated pressurized liquid extraction/liquid chromatography–mass spectrometry protocol for *Isatis tinctoria*, and qualitative analysis of other cruciferous plants. *J. Chromatogr. A.* 2007, 1166, 142–151.

48. Calbiani, F. et al. Validation of an ion-pair liquid chromatography–electrospray–tandem mass spectrometry method for the determination of heterocyclic aromatic amines in meat-based infant foods. *Food Addit. Contam.* 2007, 24, 833–841.

49. van Bommel, M.R. et al. High-performance liquid chromatography and non-destructive three-dimensional fluorescence analysis of early synthetic dyes. *J. Chromatogr. A.* 2007, 1157, 260–272.

50. Kuśmierek, K. and Bald, E. Simultaneous determination of tiopronin and d-penicillamine in human urine by liquid chromatography with ultraviolet detection. *Anal. Chim. Acta* 2007 590, 132–137.

51. Quintana, J.B. and Reemtsma, T. Rapid and sensitive determination of ethylenediamine tetraacetic acid and diethylenetriamine pentaacetic acid in water samples by ion-pair reversed-phase liquid chromatography-electrospray tandem mass spectrometry. *J. Chromatogr. A.* 2007, 1145, 110–117.

52. Mascher, D.G., Unger, C.P., and Mascher, H.J. Determination of neomycin and bacitracin in human or rabbit serum by HPLC-MS/MS. *J. Pharm. Biomed.* 2007, 43, 691–700.

53. Wybraniec, S. Effect of tetraalkylammonium salts on retention of betacyanins and decarboxylated betacyanins in ion-pair reversed-phase high-performance liquid chromatography. *J. Chromatogr. A.* 2006, 1127, 70–75.

54. Häkkinen, M.R. et al. Quantitative determination of underivatized polyamines by using isotope dilution RP-LC-ESI-MS/MS. *J. Pharm. Biomed. Anal.* 2008, 48, 414–421.

55. Gu, L., Jones, A.D., and Last, R.L. LC-MS/MS assay for protein amino acids and metabolically related compounds for large-scale screening of metabolic phenotypes. *Anal. Chem.* 2007, 79, 8067–8075.

56. Lu, C.Y. and Feng, C.H. Micro-scale analysis of aminoglycoside antibiotics in human plasma by capillary liquid chromatography and nanospray tandem mass spectrometry with column switching. *J. Chromatogr. A.* 2007, 1156, 249–253.

57. Gao, L. et al. Simultaneous quantification of malonyl-CoA and several other short-chain acyl-CoAs in animal tissues by ion-pairing reversed-phase HPLC/MS. *J. Chromatogr. B.* 2007, 853, 303–313.

58. Ansorgov, D., Holcapek, M., and Jandera, P. Ion pairing high-performance liquid chromatography-mass spectrometry of impurities and reduction products of sulphonated azo dyes. *J. Sep. Sci.* 2003, 26, 1017–1027.

59. Yokoyama, Y., Ozaki, O., and Sato, H. Separation and determination of amino acids, creatinine, bioactive amines and nucleic acid bases by dual-mode gradient ion-pair chromatography. *J. Chromatogr. A.* 1996, 739, 333–342.

60. Jaison, P.G., Raut, N.M., and Aggarwal, S.K. Direct determination of lanthanides in simulated irradiated thoria fuels using reversed-phase high-performance liquid chromatography. *J. Chromatogr. A.* 2006, 1122, 47–53.

61. Coulier, L. et al. Simultaneous quantitative analysis of metabolites using ion-pair liquid chromatography electrospray ionization mass spectrometry. *Anal. Chem.* 2006, 78, 6573–6582.

62. Ganzera, M., Lanser, C., and Stuppner, H. Simultaneous determination of *Ephedra sinica* and *Citrus aurantium* Var. amara alkaloids by ion-pair chromatography. *Talanta* 2005, 66, 889–894.

63. Chen, H.C., Wang, S.P., and Ding W.H. Determination of fluorescent whitening agents in environmental waters by solid-phase extraction and ion-pair liquid chromatography–tandem mass spectrometry. *J. Chromatogr. A.* 2006, 1102, 135–142.

64. Timmons, S.C. and Jakeman, D.L. Stereospecific synthesis of sugar-1-phosphates and their conversion to sugar nucleotides. *Carbohydr. Res.* 2008, 343, 865–874.

65. Kuśmierek, K. and Bald, E. Measurement of reduced and total mercaptamine in urine using liquid chromatography with ultraviolet detection. *Biomed. Chromatogr.* 2008, 22, 441–445.

66. Kim, J. et al. Phosphopeptide elution times in reversed-phase liquid chromatography. *J. Chromatogr. A.* 2007, 1172, 9–18.

67. Krock, B., Seguel, C.G., and Cembella, A.D. Toxin profile of *Alexandrium catenella* from the Chilean coast as determined by liquid chromatography with fluorescence detection and liquid chromatography coupled with tandem mass spectrometry. *Harm. Algae* 2007, 6, 734–744.

9 Role of Eluent pH in IPC

9.1 IPC VERSUS RP-HPLC

The retention of ionogenic analytes in RP-HPLC is obviously dependent on the eluent pH. The pH value is an important optimization parameter because it controls the extent of ionization of the solute and hence the magnitude of electrostatic interactions. Varying the acidity of the mobile phase can lead to extreme changes in selectivity. Many interdependent parameters can be modulated in IPC and their optimization requires a theory-driven procedure.

Ion pairing interactions need charged analytes to operate. Let us focus on the dependence of basic analyte retention as a function of pH. The hydrophobic retention of ionogenic bases at pH values two units above the basic analyte pK_a is the highest on reversed phase materials because the analyte is predominantly neutral. Separations at these pH levels may not be feasible because the pH stability thresholds of common silica-based stationary phases are exceeded, as discussed in Chapter 5.

Eluent pH is limited to a maximum of 7 to 8 due to the reduced chemical stability of a chromatographic bed in an alkaline medium. The nucleophilic attack of Si-O bonds by hydroxide ions leads to the erosion of the silica surface as shown by back pressure increases caused by the formation of $Si(OH)_4$. With polystyrene-divinyl-benzene-based stationary phases, pH stability is not an issue and a very wide mobile phase pH range can be used, thereby providing additional selectivity [1]. Several silica-based and polymeric columns claimed to be stable in pH ranges from 1 to 13 are commercially available, however, they are not commonly used.

In case of a weak monoprotic acid–base analyte, two different species are retained: the protonated acid form (HA, uncharged or cationic) and the deprotonated conjugate basic form (A, uncharged or anionic). In other words, if HA is uncharged, A is anionic; if HA is charged, A is uncharged. The ionization of the analyte is described by the following equilibrium:

$$HA \rightarrow H + A \tag{9.1}$$

characterized by the equilibrium constant K_a. The RP retention of this analyte, k, is:

$$k = \phi \frac{[LA]+[LHA]}{[A]+[HA]} = \phi \frac{[LA]+[LAH]}{\alpha_A * C + \alpha_{HA} * C} = \phi \frac{[LA]+[LHA]}{\frac{K_a}{[H^+]+K_a} * C + \frac{[H^+]}{[H^+]+K_a} * C} = \phi \frac{[LA]+[LHA]}{C}$$

$$k = \phi \frac{[LA]+[LHA]}{C} = \phi \frac{[LA]}{C} + \phi \frac{[LHA]}{C} = \phi \frac{[LA]}{[A]}\alpha_A + \phi \frac{[LHA]}{[HA]}\alpha_{HA}$$

$$\tag{9.2}$$

where ϕ is the ratio of the surface area of the column to the mobile phase volume (inverse phase ratio of the column), [LA] and [A] are, respectively, the surface and mobile phase concentrations of the deprotonated conjugate base, and [LHA] and [HA] are, respectively, the surface and mobile phase concentrations of the protonated acid; and α_{HA} and α_A are, respectively, the fractions of the protonated and deprotonated analyte, according to the following expressions:

$$\alpha_{HA} = \frac{[H^+]}{[H^+]+K_a} \tag{9.3}$$

$$\alpha_A = \frac{K_a}{[H^+]+K_a} \tag{9.4}$$

With decreasing eluent pH, [A] decreases and [HA] increases. Since the retention factor of the protonated and deprotonated species, k_{HA} and k_A, are respectively

$$k_{HA} = \phi\frac{[LHA]}{[HA]} \tag{9.5}$$

$$k_A = \phi\frac{[LA]}{[A]} \tag{9.6}$$

Equation 9.1 is easily demonstrated to be equivalent to the weighted sum of k_{AH} and k_A, as sanctioned by Equation 9.5 below:

$$k = \alpha_{HA}k_{HA} + \alpha_A k_A \tag{9.7}$$

When Equations 9.3 and 9.4 are inserted into Equation 9.7, the following expression can be obtained:

$$k = \frac{[H^+]k_{HA}+K_a k_A}{[H^+]+K_a} \tag{9.8}$$

that is also equivalent to:

$$k = (k_A + k_{HA}10^{\,pKa-pH})/(1 + 10^{\,pKa-pH}) \tag{9.9}$$

Equation 9.8 describes a sigmoidal decrease of reversed phase global retention of a basic analyte with decreasing pH [2].

Typical sigmoidal plots of retention of a basic analyte as a function of pH can be observed in Figure 9.1. Since HA is cationic and A is neutral for a base, $k_{HA} \ll k_A$ and retention decreases with decreasing pH. The curves are translated toward higher pH for higher pK_a of the analyte. The inflection point of the curve corresponds to

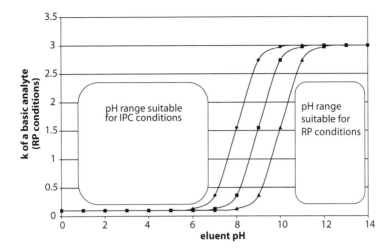

FIGURE 9.1 Theoretical curves showing basic analyte reversed phase retention as a function of pH. Diamonds: $pK_a = 10^{-8}$, Squares: $pK_a = 10^{-9}$, Triangles: $pK_a = 10^{-10}$. k_{BH+} and $k_B =$ 0.1 and 3.0, respectively.

pH $= pK_a$ of the analyte. Obviously, for acidic analytes, the mirror image of that one shown in Figure 9.1 would be theoretically obtained. Since in this case HA is neutral and A is anionic, $k_{HA} \gg k_A$ and if the pH $\ll pK_a$, the reversed phase retention would be adequate. Unfortunately, in low pH regions, the hydrophobic ligand may be hydrolyzed; hence the suppression of the acidic analyte ionization cannot be compatible with a long column lifetime. Again, also for acidic analytes, the IPC strategy is theoretically predicted (and practically demonstrated) to be an advantageous alternative to RP-HPLC.

When dealing with ionogenic analytes, it is important to bear in mind that the pH of a hydro-organic mobile phase is different from that of its aqueous fraction and the pK_a of the acid used to prepare the buffer changes with solvent composition [3–8]. Careful eluent pH control is crucial for a reproducible and successful chromatographic analysis of ionizable analytes, especially in gradient separation. A chromatographer may choose from three different pH scales. The IUPAC recommends measuring the eluent pH after mixing aqueous buffer and organic modifier. The pH electrode system can be calibrated with aqueous buffers to provide directly the $_w^s\mathrm{pH}$ values of the mobile phase [7]. This procedure is simple because it does not require pH standards for each eluent composition When the pH electrode system is calibrated with buffers prepared in a water–organic solvent mixture used as a mobile phase, the pH readings provide $_s^s\mathrm{pH}$ values [7]. The two IUPAC pH scales can be easily related according to the following relationship [8]

$$_w^s\mathrm{pH} = {}_s^s\mathrm{pH} + \delta \qquad\qquad (9.10)$$

where the δ parameter depends on the solvent used, the difference between the electrode liquid junction potentials in the eluent and water, pH standards and the type of buffer [5].

The most common pH scale used in chromatography is the aqueous $_w^w$pH scale [7]. The electrode system is calibrated with aqueous buffers and the pH is measured in the aqueous fraction of the eluent, before it is mixed with the organic modifier. This procedure is less effective than the previous methods because different buffers with the same aqueous pH can lead to significant different pH shifts after addition of the same amount of a particular organic solvent [4]. The pK_a levels of basic and acidic compounds are shifted to lower and higher values, respectively, with increasing organic modifier percentages in the eluent. This means that the neutral forms of the solute are preferred with decreasing dielectric constant of the medium [9]. The same tendency is shown by ionizable analytes. The chromatographic estimates of the analyte pK_a on the basis of the inflection point in Figure 9.1 depend on the volume fraction of the organic modifier in the eluent. Better fittings are obtained by using the $_w^s$pH or $_s^s$pH scales instead of the $_w^w$pH [4].

From Figure 9.1, it is clear that the fields (or ambit/domain…) of IPC and RP-HPLC separations related to eluent pH, are complementary; the former requires charged analytes and the latter involves their ionization suppression. The IPC of basic compounds concerns the pH region below their pK_a where they are protonated and positively charged, that matches the pH range where most commercial columns are stable. It follows that IPC is a milestone separation strategy for basic compounds. Obviously, for acidic analytes, the opposite holds and the pH should be higher than the pK_a to obtain negatively charged deprotonated species. Buffering of the mobile phase is often crucial. Phosphate, acetate, and formate buffers are the most common.

The addition of many classes of IPRs leads to a modification of the eluent pH. Volatile amines, greatly appreciated for their volatility in IPC-MS use, increase the eluent pH. Conversely, using perfluorinated acids of different chain length as IPRs for basic analytes involves a concomitant lowering of the eluent pH that in turn provides the protonation the basic analytes need under IPC conditions. The IPR counter ion is also important when dealing with eluent pH; for example, the behavior of tetrabutylammonium hydroxide would be very different from that of the corresponding chloride.

When the sigmoid curve for basic analytes was obtained by lowering the pH via the addition of $HClO_4$, CF_3COOH, and H_3PO_4 to a pH so low that molecules were completely protonated, a distortion of the typical sigmoidal plot was found because any further pH decrease led to a retention increase. Since the addition of the same amounts of the sodium salts of these acids produced the same effect on analyte retention without reducing the pH, it was concluded that the source (e.g. $HClO_4$ or $NaClO_4$) of the anions did not matter. The retention increase arose from their ion-pairing with positively charged analytes and not to the pH change [10,11]. The anion ability of acting as IPRs increases with an increase of their carbon chain for perfluorinated carboxylic acid [12].

The eluent pH is important for manipulating the analyte charge status and also for controlling the degree of ionization of the IPR that influences its ion-pairing attitude [13]. pH is also a key factor in the IPC of complexes [14,15] because it influences the ionization of the ligand and hence the degree of complex formation and redox status of the metal [16,17].

In one study, a pH gradient was exploited in the IPC of metal ions pre-complexed with a sulfonated azoligand since it affects the stability constants [14]. Simultaneous gradients of eluent pH, organic modifier, and IPR eluent concentration led to separation of several key classes of polar metabolites: very polar and acidic metabolites such as nucleotides were separated via an ion-pair mechanism while more nonpolar compounds such as CoA esters underwent a reversed phase mechanism during the same run [18] Similarly, a complicated combined organic modifier, IPR, and pH gradient allowed the baseline separation of alkaloids [19].

9.2   pH-DEPENDENT SILANOL IONIZATION

Peak tailing plagues chromatographic separations of basic compounds and may be attributed to energetic surface heterogeneity. This issue was discussed in detail in Chapter 5 (Section 5.1.1). A silica-based RP material is characterized by weak and strong adsorption sites. The former are hydrophobic alkyl chains that account for the low energy analyte hydrophobic interactions and the latter are free silanols that strongly interact with protonated compounds or electrostatically exclude anionic analytes. The presence of two distinct kinds of sites explains the bi-Langmuir adsorption equilibrium isotherms of basic compounds under simple reversed phase conditions [20,21] and leads to kinetic tailing.

The rates of mass transfer kinetics are different at different sites. The sorption–desorption of solute molecules captured by strong sites is slow compared to kinetics at the weak sites leading to an increase in band tailing [22]. Moreover, since strong sites are saturated and overloaded also at low analyte concentration, kinetic tailing (at variance with non-linear tailing) does not improve with decreasing sample loading and may exist in addition to non-linear tailing. Both types of tailing lead to asymmetrical peaks [23,24]. Therefore, flow rate and sample loading studies can determine the major origin of the tailing since kinetic tailing improves with decreasing flow rates while non-linear tailing improves with decreasing sample loading [25–27].

When dealing with silica-based RP stationary phases, consider that the pK_a values of silica hydroxyl groups (free silanols) are estimated between 5 and 7. Acidic metal impurities of the silica matrix further decrease the pK_a ranges of the silanols while new stationary phase technologies try to increase it [28]. If the mobile phase pH is below 3, most silanols would be in their neutral form and silanophilic interactions with protonated basic compounds can be ignored. In this case, hydrophobic retention is the lowest and early elution of analytes prevents good resolution unless IPC conditions are adopted. The addition of IPRs usually brings about improved efficiency and better peak shape [29–34] while providing adequate retention. Actually IPRs can shield silanophilic interactions and decrease the ion exchange interactions between the ionic compounds and the silanols also in the low pH eluents. Free silanols were considered responsible for the unexpected pH dependence of oligonucleotide retention [35].

REFERENCES

1. Toll, H. et al. Separation, detection, and identification of peptides by ion-pair reversed-phase high-performance liquid chromatography–electrospray ionization mass spectrometry at high and low pH. *J. Chromatogr. A.* 2005, 1079, 274–286.
2. Horvath, C.S., Melander, W., and Molnár, I. Liquid chromatography of ionogenic substances with nonpolar stationary phases. *Anal. Chem.* 1977, 49, 142–154.
3. Subirats, X., Bosch, E., and Rosés, M. Retention of ionisable compounds on high-performance liquid chromatography XV: estimation of pH variation of aqueous buffers with the change of the acetonitrile fraction of the mobile phase. *J. Chromatogr. A.* 2004, 1059, 33–42.
4. Subirats, X., Bosch, E., and Rosés, M. Retention of ionisable compounds on high-performance liquid chromatography XVI: Estimation of retention with acetonitrile/water mobile phases from aqueous buffer pH and analyte pKa. *J. Chromatogr. A.* 2006, 1121, 170–177.
5. Subirats, X., Bosch, E., and Rosés, M. Retention of ionisable compounds on high-performance liquid chromatography XVII: estimation of pH variation of aqueous buffers with the change of the methanol fraction of the mobile phase. *J. Chromatogr. A.* 2007, 1138, 203–215.
6. Bosch, E. et al. Retention of ionizable compounds on HPLC: pH scale in methanol–water and the pK and pH values of buffers. *Anal. Chem.* 1996, 68, 3651–3657.
7. IUPAC. *Compendium of Analytical Nomenclature: Definitive Rules*, 1997, 3rd ed. Blackwell: Oxford, 1998.
8. Espinosa, S., Bosch, E., and Rosés, M. Retention of ionizable compounds on HPLC 5: pH scales and the retention of acids and bases with acetonitrile–water mobile phases. *Anal. Chem.* 2000, 72, 5193–5200.
9. Roses, M. et al. Retention of ionizable compounds on HPLC 2: effect of pH, ionic strength, and mobile phase composition on the retention of weak acids. *Anal. Chem.* 1996, 68, 4094–4100.
10. LoBrutto, R. et al. Effect of the eluent pH and acidic modifiers in high-performance liquid chromatography retention of basic analytes. *J. Chromatogr. A.* 2001, 913, 173–187.
11. LoBrutto, R., Jones, A., and Kazakevich, Y.V. Effect of counter-anion concentration on retention in high performance liquid chromatography of protonated basic analytes. *J. Chromatogr. A.* 2001, 913, 189–196.
12. Wybraniec, S. and Mizrahi, Y. Influence of perfluorinated carboxylic acids on ion-pair reversed-phase high-performance liquid chromatographic separation of betacyanins and 17-decarboxy-betacyanins. *J. Chromatogr. A.* 2004, 1029, 97–101.
13. Szabo, Z. et al. Analysis of nitrate ion in nettle (*Urtica dioica* L.) by ion-pair chromatographic method on a C30 stationary phase. *J. Agr. Food Chem.* 2006, 54, 4082–4086.
14. Sarzanini, C. et al. Sulfonated azoligand for metal ion determination in ion interaction chromatography. *J. Chromatogr. A.* 1999, 847, 233–244.
15. Oszwaldowski S. and Pikus A. Reversed-phase liquid chromatographic simultaneous determination of iron(III) and iron(II) as complexes with 2-(5- bromo-2-pyridylazo)-5-diethylaminophenol. *Talanta* 2002, 58, 773–783.
16. Sarzanini, C. et al. Electrochemical detection of sulfonated azo dyes and their metal complexes in ion interaction chromatography. *J. Chromatogr. A.* 1998, 804, 241–248.
17. Threeprom, J., Purachaka, S., and Potipan, L. Simultaneous determination of Cr(III)-EDTA and Cr(VI) by ion interaction chromatography using a C18 column. *J. Chromatogr. A.* 2005, 1073, 291–295.
18. Coulier, L. et al. Simultaneous quantitative analysis of metabolites using ion-pair liquid chromatography–electrospray ionization mass spectrometry. *Anal. Chem.* 2006, 78, 6573–6582.

19. Ganzera, M., Lanser, C., Stuppner, H. Simultaneous determination of *Ephedra sinica* and *Citrus aurantium* Var. Amara alkaloids by ion-pair chromatography. *Talanta* 2005, 66, 889–894.
20. Quinones, I., Cavazzini, A. and Guiochon, G. Adsorption equilibria and overloaded band profiles of basic drugs in a reversed-phase system. *J. Chromatogr. A.* 2000, 877, 1–11.
21. Gritti, F. et al. Determination of single component isotherms and affinity energy distribution by chromatography *J. Chromatogr. A.* 2003, 988, 185–203.
22. Fornstedt, T., Zhong, G., and Guiochon, G. Peak tailing and slow mass transfer kinetics in nonlinear chromatography. *J. Chromatogr. A.* 1996, 742, 55–68.
23. Fornstedt, T., Zhong, G., and Guiochon, G. Peak tailing and mass transfer kinetics in linear chromatography. *J. Chromatogr. A.* 1996, 741, 1–12.
24. Gotmar, G., Fornstedt, T., and Guiochon, G. Peak tailing and mass transfer kinetics in linear chromatography: dependence on the column length and the linear velocity of the mobile phase. *J. Chromatogr. A.* 1999, 831, 17–35.
25. Buckenmaier, S.M.C., McCalley, D.V., and Euerby, M.R. Overloading study of bases using polymeric RP-HPLC columns as an aid to rationalization of overloading on silica-ODS phases. *Anal. Chem.* 2002, 74, 4672–4681.
26. McCalley, D.V. Influence of sample mass on the performance of reversed-phase columns in the analysis of strongly basic compounds by high-performance liquid chromatography. *J. Chromatogr. A.* 1998, 793, 31–46.
27. McCalley, D.V. Selection of suitable stationary phases and optimum conditions for their application in the separation of basic compounds by reversed-phase HPLC. *J. Sep. Sci.* 2003, 26, 187–200.
28. Gritti, F. and Guiochon G. Peak shapes of acids and bases under overloaded conditions in reversed-phase liquid chromatography, with weakly buffered mobile phases of various pH: a thermodynamic interpretation. *J. Chromatogr. A.* 2009, 1216, 63–78.
29. Wybraniec, S. Effect of tetraalkylammonium salts on retention of betacyanins and decarboxylated betacyanins in ion-pair reversed-phase high-performance liquid chromatography. *J. Chromatogr. A.* 2006, 1127, 70–75.
30. Roberts, J.M. et al. Influence of the Hofmeister series on the retention of amines in reversed-phase liquid chromatography. *Anal. Chem.* 2002, 74, 4927–4932.
31. Huang, Z. et al. Determination of cyclamate in foods by high performance liquid chromatography–electrospray ionization mass spectrometry. *Anal. Chim. Acta* 2006, 555, 233–237.
32. Varvaresou, A. et al. Development and validation of a reversed-phase ion-pair liquid chromatography method for the determination of magnesium ascorbyl phosphate and melatonin in cosmetic creams. *Anal. Chim. Acta* 2006, 573–574, 284–290.
33. Moret, S., Hidalgo, M., and Sánchez, J.M. Development of an ion-pairing liquid chromatography method for the determination of phenoxyacetic herbicides and their main metabolites: application to the analysis of soil samples. *Chromatographia* 2006, 63, 109–115.
34. Kawamura, K. et al. Separation of aromatic carboxylic acids using quaternary ammonium salts on reversed-phase HPLC 1: separation behavior of aromatic carboxylic acids. *Sep. Sci. Technol.* 2006, 41, 379–390.
35. McKeown, A.P., Shaw, P.N., and Barrett, D.A. Retention behaviour of an homologous series of oligodeoxythymidilic acids using reversed-phase ion-pair chromatography. *Chromatographia* 2002, 55, 271–277.

10 Temperature

Temperature is a key operational parameter that has great effects on retention, efficiency, and selectivity for IPC separations. In spite of this, column temperature was rarely considered important because results were often at variance and not easily explicable. Moreover, an easy variation of mobile phase composition (IPR concentration, organic modifier volume fraction, pH) was usually sufficient to fine tune retention, and to obtain adequate selectivity, and resolution. Quantitative and comprehensive rationalization of retention as a function of mobile phase composition [1–7] assists the chromatographer to perform educated guesses to optimize IPC separations. The lack of interest in this operational variable resulted in a few theoretical studies. While column thermostatting to improve reproducibility is becoming accepted among IPC practitioners [8], the potential of this challenging parameter remains to be explored further.

10.1 INFLUENCE OF TEMPERATURE ON COLUMN EFFICIENCY

Results of efficiency enhancement studies have been controversial. Increasing the temperature lowers eluent viscosity and system back pressure, leading to the use of (1) higher flow rates (shorter cycle times) [9], (2) longer columns, and (3) smaller particles that enhance efficiency in their own right. However, efficiency is also expected to increase because high column temperatures involve (1) faster adsorption–desorption kinetics, (2) enhanced diffusivity, (3) lower mass transfer resistance (C_m in the van Deemter Equation 6.4), and (4) flatter van Deemter curves.

All these issues explain why high temperature usually results in fast, efficient, and sufficiently rugged methods for IPC applications. Nevertheless, when efficiency was studied as a function of temperature, it was also reported to decrease after an initial increase [10]. This may probably be explained by the fact that enhanced diffusivity decreases C_m according to Equation 6.4 while increasing longitudinal diffusion (B as sanctioned by Equation 6.3) with negative effects on efficiency. Moreover, since B prevails at low eluent linear velocity while the C_m term has a stronger influence at high eluent velocity, the increase of efficiency with temperature is expected to occur mainly at high flow rates. Increased diffusivity usually improves the peak shape [10]. On the basis of these concepts, the variability of experimental results can be rationalized easily.

10.2 INFLUENCE OF TEMPERATURE ON SELECTIVITY UNDER RP–HPLC AND IPC CONDITIONS

Similarly, contrasting effects of column temperature on selectivity were reported. The following section will highlight the thermodynamic signatures of these phenomena. Thermodynamic equilibrium constants at constant pressure, at variance with stoichiometric constants, depend only on system temperature [1]. It follows that temperature is a key factor in IPC because it shapes the IPR adsorption isotherm. Because the amount of the adsorbed IPR modulates retention, temperature is a putative optimization variable. Temperature is predicted to influence selectivity since it alters the amounts of the adsorbed IPR and, at fixed surface concentration of the IPR, influences the surface potential according to Equation 3.5 (see Section 3.1.2). A temperature variation also modifies (1) ionization of buffer components [11]; (2) ionization of solutes [11]; and (3) the value of each thermodynamic equilibrium constant applicable to the chromatographic system, namely K_{LE}, K_{EH}, K_{LH}, K_{LEH}, (their meanings are clarified in Section 3.1.2 and in Reference 1) based on the general expression:

$$K = \exp(-\Delta\mu^\circ/RT) \qquad (10.1)$$

Different analytes are characterized by diverse standard free energy changes ($\Delta\mu^\circ$); thus the dependence of K_{LE} and K_{EH} on temperature is a function of analyte nature. The following Gibbs equation relates $\Delta\mu^\circ$ to ΔH° and ΔS°, that represent the standard enthalpy and entropy changes, respectively:

$$\Delta\mu^\circ = \Delta H^\circ - T\Delta S^\circ \qquad (10.2)$$

where T is absolute temperature. After substitution of Equation 10.2 into Equation 10.1, the following relationship can be easily obtained:

$$\ln K = -\Delta H^\circ/RT + \Delta S^\circ/R \qquad (10.3)$$

This expression is known as the van't Hoff equation. It is useful because the thermodynamic enthalpy and entropy parameters for the analyte transfer from the mobile to the stationary phase may be evaluated via the effect of temperature on the thermodynamic equilibrium constant. Under simple reversed phase conditions for an analyte E, K_{LE} is sufficient to describe the system and since

$$k = K_{LE}\, \phi \qquad (10.4)$$

where ϕ is the ratio of the surface area of the column to the mobile phase volume (inverse phase ratio of the column), a thermodynamic analysis of the process may be performed by recording retention of E at different temperatures and plotting $\ln k$ versus $1/T$, according to:

$$\ln k = -\Delta H^\circ/RT + \Delta S^\circ/R + \ln \phi \qquad (10.5)$$

ΔH° and ΔS° are, respectively the standard enthalpy and entropy for the analyte adsorption onto the chromatographic stationary phase. If they remain constant over the temperature range studied, the plot of Equation 10.5 should be linear, and they can be obtained, respectively, from the slope and the intercept of the plot if ϕ is known. It was, however, reported that the decrease of the dielectric constant of water with increasing temperature (due to the difficulty of water dipoles to shield electrical charges because of higher kinetic energy), its lower polarity, and the disruption of hydrogen bonds at high temperature may have altered ΔH° and ΔS°. At high

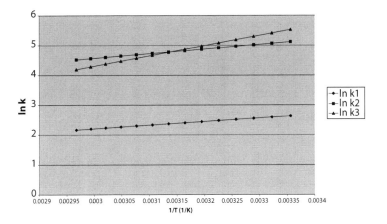

FIGURE 10.1 Typical van't Hoff plot for different model analytes.

temperatures little hydrogen bonding and solute hydration are weaker. The entropy change due to the release of hydrophobically bound water molecules upon transfer of the solute from the mobile to the stationary phase is lower, hence ΔS° is expected to decrease and not remain constant.

Conversely, at lower temperatures the hydrophobic effect entropically leads the adsorption of the solutes [12]. Actually the solvent strengths of all mixtures change with temperature [13] and this influences selectivity. Also, the non-constancy of ΔH° with changing temperature may also be due to the difference of the heat capacities of the analytes in the mobile and stationary phases according to the Kirkhoff equation [14]. Figure 10.1 is a typical van't Hoff plot for different analytes.

The temperature dependence of retention in IPC is much more complicated. As explained above, retention results from at least four interdependent equilibria, each characterized by its own ΔH° and ΔS°. The retention factor under IPC conditions (see Equation 3.19 and meanings of symbols and abbreviations in Section 3.1.2) is:

$$k = \phi \frac{[LE]+[LEH]}{[E]+[EH]} \tag{10.6}$$

that corresponds to

$$k = \phi \frac{[LE]}{[E]} \theta_E + \phi \frac{[LEH]}{[EH]} \theta_{EH} \tag{10.7}$$

where

$$\theta_E = \frac{[E]}{[E]+[EH]} = \frac{1}{1+K_{EH}[H]} \tag{10.8}$$

and

$$\theta_{EH} = \frac{[EH]}{[E]+[EH]} = 1 - \theta_E = \frac{K_{EH}[H]}{1+K_{EH}[H]} \tag{10.9}$$

If the ratio of the activity coefficient is constant, we have (see Equations 3.15 through 3.18 in Chapter 3):

$$\frac{[LE]}{[E]} = K_{LE} \exp\left(-\frac{z_E F \Psi^\circ}{RT}\right)[L] \tag{10.10}$$

and

$$\frac{[LEH]}{[EH]} = \frac{K_{LEH}}{K_{EH}}[L] \tag{10.11}$$

From the substitution of Equations 10.10 and 10.11 in Equation 10.7, the following expression can be easily obtained:

$$k = \phi[L]\left(K_{LE} \exp\left(-\frac{z_E F \Psi^\circ}{RT}\right)\theta_E + \frac{K_{LEH}}{K_{EH}}\theta_{EH}\right) \tag{10.12}$$

Equation 10.12 is algebraically correspondent [15] to the final relationship that describes analyte retention under IPC conditions (Equation 3.21). It upgrades the parallel stoichiometric equation of the model by Kazakevich and co-workers [16] that is inherently inadequate because it cannot predict the decrease of retention for analytes similarly charged to the chaotropic reagent and the electrostatic tuning of the retention of the unpaired analyte in the presence of the electrified stationary phase. It also upgrades electrostatic models [17,18] that disregard the role played by the ion-pair complex (final term in Equation 10.12).

Equation 10.12 is more informative than the equivalent Equation 3.21 for interpreting the influence of temperature under IPC conditions. Equation 10.12 indicates that the experimental capacity factor is actually the weighted average of the electrostatically modulated retention factors of the free analyte and that of the paired analyte. Notably, the global ΔH° for IPC retention can be thought of as a weighted average of the retention ΔH° of both free (electrostatically tuned) analyte and paired analyte. This global ΔH° can be calculated by differentiating the logarithm of Equation 10.6 with respect to (1/T). The weighted averaged ΔH° was tentatively calculated only on the basis of a stoichiometric approach and under the additional assumption that ion-pair complex formation in the mobile phase is the dominant process [19]. The conclusions, however, are dubious.

We reinforce the concept that this approach is not acceptable because it does not account for the charge density of the stationary phase verified by extensive experimental evidence. The disregarded electrostatic potential that develops depends on column temperature, as explained above. Hence a sound theoretical description of the influence of temperature on retention enthalpy is still needed. The concept of the weighted averaged $\Delta H°$ may explain why, even if solute exists in two forms in the mobile phase (each with its own retention factor) and undergoes both hydrophobic and electrostatic interactions with the stationary phase, a linear van't Hoff plot is usually obtained also under IPC conditions [20], at least for a limited temperature range.

Examples of linear van't Hoff relationships can be found in the literature. Almost linear relationships between ln$k$ versus 1/T were reported for alkaloids under IPC conditions at a temperature range of 10 to 50°C with a linear decrease of retention with the increase of temperature [10]. A linear van't Hoff plot was also obtained in IPC of impurities in 2,4-disulfonic acid benzaldehyde di-sodium salt [21], even if nonlinear dependencies were also reported [22].

When the van't Hoff lines for different analytes are not parallel, a temperature increase is theoretically predicted to increase or decrease selectivity when the lines diverge or converge with increasing temperature. Figure 10.1 illustrates this concept. An inversion of selectivity and peak crossover is also possible [23,24]. Even if the selectivity does not improve because the van't Hoff lines are parallel, increased efficiency and lower retention may be valuable consequences when a chromatographic run is performed at higher temperature [21]. Actually a fast IPC of urinary thiocyanate benefits from a temperature of 45°C [25]. Similarly in IPC of underivatized carboxylic acids on a porous graphitic stationary phase, retention was lower at higher column temperature even if the analyte behavior toward temperature was variable and the separation was better at low temperature [26].

Since selectivity is a prerequisite for resolution, a temperature increase strongly impacts both selectivity and resolution. The variability of the van't Hoff lines explains the seemingly erratic result of the influence of temperature on the figures of merit of the method. For example, at lower temperatures, the resolution of model peptides with +1 and +3 net charges improved and worsened, respectively [27]. In chiral IPC, T < 0°C was successfully used to improve enantioresolution [28]. Similarly, lower temperatures provided better resolution in the analysis of a new aminoglycoside antibiotic [29] and for characterization of maize products [30]. Conversely, an increased resolution at 70°C was observed when the ion-pair mechanism was exploited under IPC–capillary zone electrophoresis of cationic proteomic peptide standards [31].

Interestingly the temperature was a crucial parameter to achieve DNA [32–35] and RNA denaturation [36] (see DHPLC in Chapter 16 for further details).

The column temperature was also an important factor to successfully join IPC with atomic absorption spectrometry in the online speciation of Cr(III) and Cr(VI) [37]. Some concerns for high temperature methods include column stability [38] and potential on-column degradation of analytes [39].

REFERENCES

1. Cecchi, T., Pucciarelli, F., and Passamonti, P. Extended thermodynamic approach to ion-interaction chromatography. *Anal. Chem.* 2001, 73, 2632–2639.
2. Cecchi, T. Extended thermodynamic approach to ion-interaction chromatography: a thorough comparison with the electrostatic approach and further quantitative validation. *J. Chromatogr. A.* 2002, 958, 51–58.

3. Cecchi, T., Pucciarelli, F., and Passamonti, P. Ion interaction chromatography of neutral molecules. *Chromatographia* 2000, 53, 27–34.

4. Cecchi, T. et al. Dipole approach to ion interaction chromatography of zwitterions. *Chromatographia* 2001, 54, 38–44.

5. Cecchi, T., Pucciarelli, F., and Passamonti, P. Extended thermodynamic approach to ion-interaction chromatography: influence of the organic modifier concentration. *Chromatographia* 2003, 58, 411–419.

6. Cecchi, T., Pucciarelli, F., and Passamonti P. Extended thermodynamic approach to ion interaction chromatography: a mono- and bivariate strategy to model the influence of ionic strength. *J. Sep. Sci.* 2004, 27, 1323–1332.

7. Cecchi, T., Pucciarelli, F., and Passamonti, P. Ion interaction chromatography of zwitterions: fractional charge approach to model the influence of the mobile phase concentration of the ion interaction reagent. *Analyst* 2004, 129, 1037–1042.

8. Vaněrková, D., Jandera, P., and Hrabica, J. Behaviour of sulphonated azo dyes in ion-pairing reversed-phase high-performance liquid chromatography. *J. Chromatogr. A.* 2007, 1143, 112–120.

9. Trojer, L. et al. Monolithic poly(p-methylstyrene-co-1,2-bis(p-vinylphenyl)ethane) capillary columns as novel styrene stationary phases for biopolymer separation. *J. Chromatogr. A.* 2006, 1117, 56–66.

10. Waksmundzka-Hajnos, M., Petruczynik, A., and Ciesśla, G. Temperature: the tool in separation of alkaloids by RP-HPLC. *J. Liq. Chromatogr. Rel. Technol.* 2007, 30, 2473–2484.

11. Heinisch, S. et al. Effect of temperature on the retention of ionizable compounds in reversed-phase liquid chromatography: application to method development. *J. Chromatogr. A.* 2006, 1118, 234–243.

12. Coym, J.W. and Dorsey, J.G. Reversed-phase retention thermodynamics of pure-water mobile phases at ambient and elevated temperature. *J. Chromatogr. A.* 2004, 1035, 23–29.

13. Colin, H. et al. The role of temperature in reversed-phase high performance liquid chromatography using pyrocarbon-containing adsorbents. *J. Chromatogr.* 1978, 166, 41–65.

14. Wieprecht, T. et al. Role of helix formation for the retention of peptides in reversed phase high-performance liquid chromatography. *J. Chromatogr. A.* 2001, 912, 1–12.

15. Cecchi, T. and Passamonti, P. Retention mechanism for ion-pair chromatography with chaotropic additives. *Chromatogr. A.* 2009, 1216, 1789–1797 and Erratum in *J. Chromatgr. A.* 2009, 1216, 5164.

16. LoBrutto, R., Jones, A., and Kazakevich, Y.V. Effect of counter-anion concentration on retention in high-performance liquid chromatography of protonated basic analytes. *J. Chromatogr. A.* 2001, 913, 189–196.

17. Bartha, A. and Ståhlberg, J. Electrostatic retention model of reversed-phase ion-pair chromatography. *J. Chromatogr. A.* 1994, 668, 255–284.

18. Cantwell, F.F. Retention model for ion-pair chromatography based on double-layer ionic adsorption and exchange. *J. Pharm. Biomed. Anal.* 1984, 2, 153–164.

19. Li, J. Effect of temperature on selectivity in reversed-phase liquid chromatography: a thermodynamic analysis. *Anal. Chim. Acta* 1998, 369, 21–37.

20. Gennaro, M.C. et al. Temperature dependence of retention in reversed-phase ion-interaction chromatography. *J. Chromatogr. Sci.* 1995, 33, 360–364.

21. Lu, B. Jonsson, P., and Blomberg, S. Reversed phase ion-pair high performance liquid chromatographic gradient separation of related impurities in 2,4-disulfonic acid benzaldehyde di-sodium salt. *J. Chromatogr. A.* 2006, 1119, 270–276.

22. Luo, X. et al. Analysis of 3′, 5′ reversed-sequence oligonucleotide isomers by reversed-phase ion-pair chromatography *Chin. J. Chromatogr.* 2007, 25, 814–819.

23. Borch, T. and Gerlach, R. Use of reversed-phase high-performance liquid chromatography–diode array detection for complete separation of 2,4,6-trinitrotoluene metabolites and EPA Method 8330 explosives: influence of temperature and an ion-pair reagent. *J. Chromatogr. A.* 2004, 1022, 83–94.

24. Miwa, H. and Yamamoto, M. Determination of mono-, poly- and hydroxy-carboxylic acid profiles of beverages as their 2-nitrophenylhydrazides by reversed-phase ion-pair chromatography. *J. Chromatogr. A.* 1996, 721, 261–268.

25. Connolly, D. Barron, L., and Paull, B. Determination of urinary thiocyanate and nitrate using fast ion-interaction chromatography. *J. Chromatogr. B.* 2002, 767, 175–180.

26. Chaimbault, P. et al. Ion-pair chromatography on a porous graphitic carbon stationary phase for the analysis of twenty underivatized protein amino acids. *J. Chromatogr. A.* 2000, 870, 245–254.

27. Chen, Y. et al. Optimum concentration of trifluoroacetic acid for reversed-phase liquid chromatography of peptides revisited. *J. Chromatogr. A.* 2004, 1043, 9–18.

28. Karlsson, A. and Charron, C. Reversed-phase chiral ion-pair chromatography at a column temperature below 0°C using three generations of Hypercarb as solid-phase. *J. Chromatogr. A.* 1996, 732, 245–253.

29. Xi, L., Wu, G., and Zhu, Y. Analysis of etimicin sulfate by liquid chromatography with pulsed amperometric detection. *J. Chromatogr. A.* 2006, 1115, 202–207.

30. Rodriguez-Nogales, J.M., Garcia, M.C., and Marina, M.L. Development of a perfusion reversed-phase high performance liquid chromatography method for the characterisation of maize products using multivariate analysis. *J. Chromatogr. A.* 2006, 1104, 91–99.

31. Popa, T.V., Mant, C.T., and Hodges, R.S. Ion interaction–capillary zone electrophoresis of cationic proteomic peptide standards. *J. Chromatogr. A.* 2006, 1111, 192–199.

32. Oberacher, H. et al. Direct molecular haplotyping of multiple polymorphisms within exon 4 of the human catechol-O-methyltransferase gene by liquid chromatography–electrospray ionization time-of-flight mass spectrometry. *Anal. Bioanal. Chem.* 2006, 386, 83–91.

33. Narayanaswami, G. and Taylor, P.D. Site-directed mutagenesis of exon 5 of p53: purification, analysis, and validation of amplicons for DHPLC. *Genet. Test.* 2002, 6, 177–184.

34. Pavlova, A. et al. Detection of heterozygous large deletions in the antithrombin gene using multiplex polymerase chain reaction and denatured high performance liquid chromatography. *Haematologica* 2006, 91, 1264–1267.

35. Shi, R. et al. Temperature-mediated heteroduplex analysis for the detection of drug-resistant gene mutations in clinical isolates of *Mycobacterium tuberculosis* by denaturing HPLC, SURVEYOR nuclease. *Microb. Infect.* 2006, 8, 128–135.

36. Beverly, M. et al. Liquid chromatography electrospray ionization mass spectrometry analysis of the ocular metabolites from a short interfering RNA duplex. *J. Chromatogr. B.* 2006, 835, 62–70.

37. Sarica, D.Y., Turker, A.R., and Erol, E. On-line speciation and determination of Cr(III) and Cr(VI) in drinking and waste water samples by reversed-phase high performance liquid chromatography coupled with atomic absorption spectrometry. *J. Sep. Sci.* 2006, 29, 1600–1606.

38. Nawrocki, J. et al. Part II. Chromatography using ultrastable metal oxide-based stationary phases for HPLC. *J. Chromatogr. A.* 2004, 1028, 31–62.

39. Xu, Q.A. and Trissel, L.A. *Stability-Indicating HPLC Methods for Drug Analysis.* Pharmaceutical Press: London, 1999.

11 Special IPC Modes and Variations

Recently many variants of typical IPC set-ups were explored. The examples described below demonstrate the broadness of scope of this finely tunable separation strategy.

11.1 MIXTURE OF DIFFERENT IPRs IN MOBILE PHASE

During a typical IPC separation, a single IPR is present in the mobile phase, usually at constant concentration. The possibilities offered by the simultaneous presence of more than one hydrophobic ion in the mobile phase were recently investigated. An atypical concurrent use of both anionic and cationic IPRs yielded more efficiency and a shorter analysis time, probably because of the competition between the solute and the similarly charged IPR [1,2]. The same strategy adopted in the IPC separation of oxytetracycline and its marker residue in edible animal tissues was optimized using an excess of IPR oppositely charged to the analytes [3].

Ionic liquids (ILs), introduced to suppress deleterious effects of silanophilic interactions, soon proved to be extraordinary IPRs (see Chapter 7.4 for details). An ILs is an equimolar mixture of cationic and anionic hydrophobic ions, each one able to adsorb onto the stationary phase. The synergistic contributions of both cationic and anionic components generates the unique properties of ILs.

The electrified stationary phase carries the same charge status of the IL ion that shows the strongest adsorbophilic attitude. Furthermore, ionic interactions between the analyte ion and the IL anion and cation, respectively, are contradictory and concur to modulate analyte ion retention in a complicated way. It follows that by increasing IL in the eluent, overall retention of the analyte may potentially (1) decrease [4] or (2) increase [5,6], or (3) remain almost constant if the conflicting effects of the IL cation and anion balance each other [7], depending on the specific IL concentration in the mobile phase [8]. Furthermore a reversal of elution sequence with increasing IL concentration is possible [9]. The multiplicity of interactions in the presence of a mixture of these ionic modifiers offers wide versatility related to selectivity adjustment.

A mixture of zwitterionic and cationic surfactants yielded superb separation of common anions on a C18 column, while a mixture of a zwitterionic and anionic modifiers proved valuable in the separation of carboxylic acids under reversed phase conditions. It reduced the chromatographic run time and improved peak shape [10].

11.2 PERMANENTLY COATED COLUMNS

Strong adsorption of many classical IPRs (long chain organic nitrogen cations and long chain sulfates and sulfonates) onto chromatographic stationary phase packings is usually a drawback of IPC. The original column performance may be difficult to restore because these IPRs tend to stick very strongly to the stationary phase. Their desorption is not an easy process and may also be incomplete even if many void volumes of hydro-organic mixture are eluted.

Increased column temperature and the presence of small amounts of salts to provide adsorbed IPR counter ions may be helpful. Despite these efforts, durable adsorption may translate to a favorable attribute of long chain IPRs, since it is conceivable to cover the stationary phase with the lipophilic surfactant and have it mimic the role of an ion exchanger when IPR presence in the eluent is discontinued. Since the stationary phase chemistry controls the selectivity of the separation, simply coating the same stationary phase with a different surfactant or mixture of surfactants is a straightforward way of adjusting separation selectivity.

Anion exchangers [11,12], cation exchangers [13], and zwitterionic phases [14] have been created and successfully tested for fast ion chromatography [12]. For example, fast ion chromatography of a common anion was accomplished by coating the stationary phase with very hydrophobic didodecyldimethylammonium bromide every 3 to 5 days and eliminating the IPR from the eluent. The separation was performed at high pH since a bidentate linkage of the bonded C18 ligands avoided silica alkaline dissolution [11].

To avoid basic silica dissolution that leads to poor reproducibility, reduced efficiencies, poor peak shapes and high back pressure carbon-based stationary phases were also tested. Unlike silica-based reversed phases, they can be exposed to both highly acidic and alkaline environments (pH 1 to 14) and very high temperatures without degradation [15]; they proved very useful in the analysis of organic acids [16].

At variance with classical ion exchangers, column anion exchange capacity can be tuned easily by adjusting the organic modifier content of the IPR coating solution; this represents an advantage of this approach compared to classical ion chromatography. Recently the perceived lack of stability of surfactant coatings that hinders their use for routine separations was studied under optimizing coating conditions (organic modifier content, ionic strength, surfactant concentration, and temperature) to obtain the best column stability and control of column capacity.

Optimum results can be obtained if the surfactant in the coating solution is below its critical micelle concentration since in this case the column is stable for ≥3000 volumes [17]. Permanent coating of the stationary phase is a cost-effective strategy, since it needs only minimal IPR. Conversely, in typical IPC, a dynamic coating of the stationary phase is obtained via a continuous flow of IPR-containing eluent; hence large amounts of reagents are needed. Moreover, permanent coating of the stationary phase elutes the analyte free from the ionic modifier. This strategy also was effective in the gradient separation of the entire lanthanide series in a fission product mixture on a reverse phase column coated with di-(2-ethylhexyl) phosphoric acid—an excellent extractant for lanthanides [18].

Permanently coated stationary phases were also used in the presence of an IPR in the eluent different from that used for the coverage. Incorporation of a non-ionic surfactant in the coating solution proved very useful for improving peak shape; the underlying material in coated columns still participates in analyte retention [10].

11.3 IPRs ADDED ONLY TO SAMPLE SOLUTION AND GHOST PEAKS

The use of permanently coated columns was not the only strategy to eliminate IPRs in eluents. The possibility of adding an IPR only to the solution of sample to be injected was successfully explored. It was suggested that ion-pairing in a sample is sufficient to provide improved reversed phase retention. The effectiveness of this stratagem has both theoretical and practical value.

It definitively demonstrates that ion-pairing among hydrophobic ions occurs in water—at variance with the Bjerrum's electrostatic modeling of ion-pairing (see Chapter 2.5.3 for further details); impairs the foundations of electrostatic retention models that neglect the importance of ion-pairing in the eluent [19,20]; and confirms the need to treat the ion-pair formation at a thermodynamic level [21,22].

For example, in the determination of N-nitrosodiethanolamine in cosmetic products, sodium 1-octanesulfonate served as the IPR; it formed a complex with the analyte that was subsequently analyzed under reversed phase conditions. The best performance was obtained for a 1:1 molar ratio between the analyte and the IPR; a molar excess of the IPR did not improve it. The ion-pair formation between the analyte and the IPR was definitively confirmed by negative ion electrospray ionization mass spectrometry. The MS/MS measurements of the resulting fragments confirmed the presence of the octanesulfonate ion in the complex [23]. The same strategy was used for the selective enrichment of glycopeptides from glycoprotein digests; bizarrely positively charged peptides were paired with chloride ion, a very atypical IPR, that was added only to the injection solution. Peptide molecules converted into their neutral forms by ion-pairing became more hydrophobic [24].

Note that co-elution of the analyte and IPR in the form of an ion-pair is not a rule. A dynamic distribution equilibrium of both the IPR and analyte between the plug of injected sample and the stationary phase may also involve a separation of the ion-pair partners if their retention free energies are very different. Moreover, since the hydrophobicities of the analyte and the ion-pair between analyte and IPR differ, a split, broad, or asymmetric peak may also be observed. This happens if the rate of interconversion between the free and paired analyte is slow compared to the chromatographic retention time scale and this downside can be observed in Figure 11.1 [25]. In this case, the analyte–IPR ion-pair would not be detected via MS [26]. Interestingly, analyte retention increased with the alkyl chains of the IPR in the reconstitution solution, similar to traditional IPC [26].

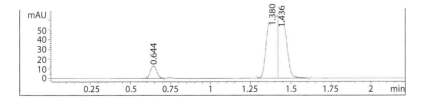

FIGURE 11.1 Split peak (1.380–1.436) of interconverting free and paired analytes if the rate of the process is slow compared to chromatographic retention timescale.

This newly introduced strategy is inexpensive because very low amounts of IPR are consumed; satisfactory retention can be achieved without long column equilibration times. Moreover, when IPC is combined with MS, the absence of IPR in the eluent increases the sensitivity of the method because the ionization suppression in the MS source due to ion-pairing is lower [27].

Note that it is a common practice to dissolve the analytes to be separated in the mobile phase. In fact, the injection of the sample in a solution whose composition differs from that of the eluent would create a disturbance at the top of the column that will migrate down the column. The lack of detectable eluent component in the injected solution would create a negative system, perturbation, or "ghost" peak because the concentration of that component is lower compared to the initial conditions under which the baseline was recorded. Obviously the retention time of the system peak is that of the component under the experimental conditions used in the chromatographic run [28,29]. System peaks are also very useful for indirect detection [30], as will be explained in Chapter 13.

11.4   SPECIAL ADDITIVES IN MOBILE PHASE

The peak shapes of metal chelating analytes are often poor because metal impurities in the stationary phase behave as active sites characterized by slower desorption kinetics and higher interaction energies compared to reversed phase ligand sites. This phenomenon is typical of silica-based stationary phases [31]; ultrapure silicas were made commercially available to reduce it. However, styrene-divinylbenzene-based chromatographic packings suffer from the same problem and it was hypothesized that metals may be present in the matrix at trace conditions because they were used as additives in the polymerization process; they may have been captured via Lewis acid–base interactions between the aromatic ring π electrons and impurities in the mobile phase [32].

The peak distortion in the IPC separation of oxytetracycline and its marker residue in edible animal tissues, probably due to their chelating abilities, was totally eliminated via the addition of EDTA to the mobile phase [3]. The chelating ability of EDTA that masks heavy metal ions was profitably used for IPC separation of all isomers of pyridinedicarboxylic acids [33]. EDTA was also used to prevent the masking of nitrite peaks by excess iron [34], and was also useful for avoiding the disruption of ion-pair retention mechanisms by calcium ions in the IPC determination of risedronate [35]. The main downside of EDTA is that its removal from the stationary phase is very difficult. Potassium tetrakis (1H-pyrazolyl) borate proved superior to EDTA for improving the peak shapes of chelating analytes and because it is removed more easily from a chromatographic column and may be used in extreme pH conditions [36]. Various silanol screening reagents including ILs can be successfully used to improve peak shapes [37] (see Section 5.1.1 for further details).

11.5   MODIFICATION OF ELUENT IONIC STRENGTH
AT CONSTANT IPR CONCENTRATION

It is a common practice in IPC to buffer the mobile phase at a suitable pH to control analyte ionization and avoid analyte acid–base equilibrium that would be deleterious for peak shape. For a long time, inorganic components of a buffer were considered "indifferent electrolytes" compared to classical hydrophobic IPRs. The breakthrough of chaotropic inorganic ions as IPRs questioned the validity of this presumption, as explained in Chapter 7. The role of each ionic component of the mobile phase deserves consideration because all ionic components exert electrostatic interactions with the sample ions. Even a salt considered neutral in the Hofmeister series, namely NaCl, was demonstrated to ion-pair successfully [24,38]. Increasing the eluent concentration of an ionic modifier (IM) less adsorbophilic than the IPR ions results in [39]:

1. Lower electrostatic surface potential according to Equation 3.5 because the fixed surface charges are better shielded by the IM ions in the electrical double layer; this promotes analyte elution via a competition between IM and analyte ions for the adsorbed IPR.
2. A Donnan effect that prevents self-repulsion of the similarly charged adsorbed IPR ions; at higher ionic strength, the higher surface concentration of the IPR ($[LH]$ in Equation 3.5) partially compensates for the decreased magnitude of the electrostatic surface potential due to the increased Σc_{0i} in Equation 3.5.
3. Salting-in or salting-out of analytes as a consequence of the addition of the IM that leads to decreased or increased surface tension, respectively, of the eluent. IPC ions are usually salting-in agents because they lower the surface tension of the eluent [40]; it was recently demonstrated that these salting effects may be revised based on the ion-pair concept.

As a result of the interlocking issues cited above, IPC retention usually varies with increasing ionic strength via the addition of an IM [39,41–43], according to the analyte charge [42,44,45]. These effects can be observed in Figure 11.2 which details the retention behavior of positively (E$^+$) and negatively (E$^-$) charged analytes in the presence

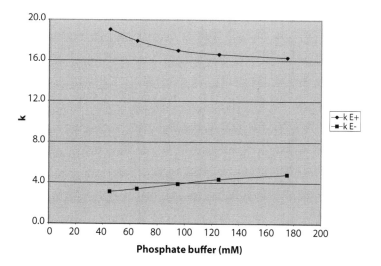

FIGURE 11.2 Retention of a positively (E$^+$) and negatively (E$^-$) charged analytes in the presence of fixed amount of negatively charged IPR as a function of the buffer ionic strength.

TABLE 11.1

Influences of Most Important Factors on Analyte Retention in Ion-Pair Chromatography

	Absolute Value of Analyte Charge	Increased Stationary Phase Lipophilicity	Increased Stationary Phase Surface Area	Increased Organic Modifier Concentration	Increased IPR Concentration	Increased Chain Length or Hydrophobicity of the IPR	pH Change	Temperature Increase
Effect on retention of analyte oppositely charged to IPR	Amplifies effects of other parameters	Retention increases	Retention increases	Retention decreases more than in absence of IPR	Retention increases up to threshold value; retention maxima can also be observed	Retention increases	Retention increases when pH maximizes analyte ionization	Retention decreases
Effect on retention of analyte similarly charged to IPR	Amplifies effects of other parameters	Retention increases	Retention increases	Retention decreases less than in absence of IPR	Retention decreases	Retention decreases	Retention decreases when pH maximizes analyte ionization	Retention decreases

of a fixed amount of a negatively charged IPR and increasing buffer ionic strength. Clearly, the shield effect overwhelms the others and the lower electrostatic potential (negative, same charge status of the IPR) results in a decreased (increased) retention for the oppositely (similarly) charged analytes. Even when IMs did not greatly affect analyte retention, they appeared functional for improving peak shape [42].

When a stepwise gradient (sudden switch between two mobile phases) elution changed the solvent strength more quickly than a continuous gradient, decreasing only the phosphate buffer concentration and keeping eluent IPR and organic modifier concentration constant, a peak crossover was detected because one solute previously moving slower than another during gradient elution migrated faster and reached the column outlet earlier [46].

Table 11.1 summarizes the influences of the most important factors on analyte retention in IPC and condenses the concept expressed in Chapters 3 to 11.

REFERENCES

1. Yalcın, G. and Yuktas, F.N. An efficient separation and method development for the quantifying of two basic impurities of nicergoline by phase high performance liquid chromatography using ion-pairing counter ions. *J. Pharm. Biomed. Anal.* 2006, 42, 434–440.
2. Fletouris, D.J. et al. Highly sensitive ion-pair liquid chromatographic determination of albendazole marker residue in animal tissues. *J. Agr. Food Chem.* 2005, 53, 893–898.
3. Fletouris, D.J. and Papapanagiotou, E.P. A new liquid chromatographic method for routine determination of oxytetracycline marker residue in the edible tissues of farm animals. *Anal. Bioanal. Chem.* 2008 391, 1189–1198.
4. He, L. et al. Effect of 1-alkyl-3-methylimidazolium-based ionic liquids as the eluent on the separation of ephedrines by liquid chromatography. *J. Chromatogr. A.* 2003, 1007, 39–45.
5. Berthod, A., Ruiz-Angel, M.J., and Huguet, S. Nonmolecular solvents in separation methods: dual nature of room temperature ionic liquids. *Anal. Chem.* 2005, 77, 4071–4080.
6. Tang, F. et al. Determination of octopamine, synephrine and tyramine in citrus herbs by ionic liquid improved 'green' chromatography. *J. Chromatogr. A.* 2006, 1125, 182–188.
7. Ruiz-Angel, M.J., Carda-Broch, S., and Berthod A. Ionic liquids versus triethylamine as mobile phase additives in the analysis of β-blockers. *J. Chromatogr. A.* 2006, 1119, 202–208.
8. Jin, C.H., Polyakova, Y., and Kyung, H.R. Effect of concentration of ionic liquids on resolution of nucleotides in reversed-phase liquid chromatography. *Bull. Korean Chem. Soc.* 2007 28, 601–606.
9. Yoo, C.G., Han, Q., and Mun, S. Reversal of elution sequence and selectivity resulting from the use of an ionic liquid as a mobile phase modifier. *J. Liq. Chromatogr. Rel. Technol.* 2008, 31, 1104–1122.
10. Fritz, J.S., Yan, Z., and Haddad, P.R. Modification of ion chromatographic separations by ionic and nonionic surfactants. *J. Chromatogr. A.* 2003, 997, 21–31.
11. Pelletier, S. and Lucy C.A. Fast and high-resolution ion chromatography at high pH on short columns packed with 1.8 μm surfactant-coated silica reverse-phase particles *J. Chromatogr. A.* 2006, 1125, 189–194.

12. Connolly, D. and Paull, B. Fast ion chromatography of common inorganic anions on a short ODS column permanently coated with didodecyldimethylammonium bromide. *J. Chromatogr. A.* 2002, 953, 299–303.

13. Gillespie, E. et al. Evaluation of capillary ion exchange stationary phase coating distribution and stability using radial capillary column contactless conductivity detection. *Analyst* 2006, 131, 886–888.

14. Riordan, C.O. et al. Double gradient ion chromatography using short monolithic columns modified with a long chained zwitterionic carboxybetaine surfactant. *J. Chromatogr. A.* 2006, 1109, 111–119.

15. Chambers, S.D. and Lucy C.A. Surfactant-coated graphitic carbon-based stationary phases for anion-exchange chromatography. *J. Chromatogr. A.* 2007, 1176, 178–184.

16. Yoshikawa, K. et al. *Talanta.* Ion chromatographic determination of organic acids in food samples using a permanent coating graphite carbon column. 2007, 72, 305–309.

17. Glenn, K.M. and Lucy, C.A. Stability of surfactant coated columns for ion chromatography. *Analyst* 2008, 133, 1581–1586.

18. Sivaraman, N. et al. Separation of lanthanides using ion-interaction chromatography with HDEHP coated columns. *J. Radioanal. Nucl. Chem.* 2002, 252 , 491–495.

19. Bartha, A. and Ståhlberg, J. Electrostatic retention model of reversed-phase ion-pair chromatography. *J. Chromatogr. A.* 1994, 668, 255–284.

20. Cantwell, F.F. Retention model for ion-pair chromatography based on double-layer ionic adsorption and exchange. *J. Pharm. Biomed. Anal.* 1984, 2, 153–164.

21. Cecchi, T., Pucciarelli, F., and Passamonti, P. Extended thermodynamic approach to ion-interaction chromatography. *Anal. Chem.* 2001, 73, 2632–2639.

22. Cecchi, T. Extended thermodynamic approach to ion-interaction chromatography: a thorough comparison with the electrostatic approach and further quantitative validation. *J. Chromatogr. A.* 2002, 958, 51–58.

23. Ghassempour, A. et al. Monitoring of N-nitrosodiethanolamine in cosmetic products by ion-pair complex liquid chromatography and identification with negative ion electrospray ionization mass spectrometry. *J. Chromatogr. A.* 2008, 1185, 43–48.

24. Ding, W., Hill, J.J., and Kelly, J. Selective enrichment of glycopeptides from glycoprotein digests using ion-pairing normal-phase liquid chromatography. *Anal. Chem.* 2007, 79, 8891–8899.

25. Keller, R.A. and Giddings, J.C. Multiple zones and spots in chromatography. *J. Chromatogr.* 1960, 3, 205–220.

26. Gao, S. et al. Evaluation of volatile ion-pair reagents for the liquid chromatography–mass spectrometry analysis of polar compounds and application to the determination of methadone in human plasma. *J. Pharm. Biomed. Anal.* 2006, 40, 679–688.

27. Marín, J.M. et al. An ion-pairing liquid chromatography/tandem mass spectrometric method for the determination of ethephon residues in vegetables. *Rapid Commun. Mass Spectrom.* 2006, 20, 419–426.

28. Gennaro, M.C. Ghost, vacancy, dip, or system peaks? A contribution in investigations about injection and system peaks in liquid chromatography. *J. Liq. Chromatogr. Rel. Technol.* 1987, 10, 3347–3375.

29. Bidlingmeyer, B.A. et al. Retention mechanism for reversed-phase ion-pair liquid chromatography. *J. Chromatogr.* 1979, 186, 419–434.

30. Forssén, P. and Fornstedt, T. General theory of indirect detection in chromatography. *J. Chromatogr. A.* 2006, 1126, 268–275.

31. Cecchi, T. et al. Influence of metal impurities sorption onto a silica based C18 stationary phase on the HPLC of metal chelating analytes. *J. Liq. Chromatogr. Rel. Technol.* 1999, 22, 429–440.

32. Cecchi, T., Pucciarelli, F., and Passamonti, P. Influence of metal ions sorption onto a styrene–divinylbenzene C18 stationary phase on the HPLC of metal chelating analytes. *J. Liq. Chromatogr. Rel. Technol.* 1999, 22, 2467–2481.

33. Pucciarelli, F., Passamonti, P., and Cecchi, T. Separation of all isomers of pyridinedicarboxylic acids by ion-pairing chromatography. *J. Liq. Chromatogr. Rel. Technol.* 1997, 20, 2233–2240.

34. Butt, S.B, Riaz, M., and Iqbal, M.Z. Simultaneous determination of nitrite and nitrate by normal phase ion-pair liquid chromatography. *Talanta* 2001, 55, 789–797.

35. Jia, H.J., Li, W., and Zhao. K. Determination of risedronate in rat plasma samples by ion-pair high-performance liquid chromatography with UV detector. *Anal. Chim. Acta* 2006, 562, 171–175.

36. Cecchi, T., Pucciarelli, F., and Passamonti, P. Potassium tetrakis (1H-pyrazolyl)borate: a mobile phase additive for improved chromatography of metal chelating analytes. *J. Liq. Chromatogr. Rel. Technol.* 1997, 20, 3329–3337.

37. Reta, M. and Carr, P.W. Comparative study of divalent metals and amines as silanol-blocking agents in reversed-phase liquid chromatography. *J. Chromatogr. A.* 1999, 855, 121–127.

38. Cecchi, T. and Passamonti, P. Retention mechanism for ion-pair chromatography with chaotropic additives. *J. Chromatogr. A.* 2009, 1216, 1789–1797 and *Erratum in J. Chromatogr. A.* 2003, 1216–5164.

39. Cecchi, T., Pucciarelli, F., and Passamonti, P. Extended thermodynamic approach to ion interaction chromatography: a mono- and bivariate strategy to model the influence of ionic strength. *J. Sep. Sci.* 2004, 27, 1323–1332.

40. Horvath, C. et al. Enhancement of retention by ion-pair formation in liquid chromatography with nonpolar stationary phases. *Anal. Chem.* 1977, 49, 2295–2305.

41. Lu, B., Jonsson, P., and Blomberg, S. Reversed phase ion-pair high performance liquid chromatographic gradient separation of related impurities in 2,4-disulfonic acid benzaldehyde di-sodium salt. *J. Chromatogr. A.* 2006, 1119, 270–276.

42. Bruzzoniti, M.C., Mentasti, E., and Sarzanini, C. Divalent pairing ion for ion-interaction chromatography of sulfonates and carboxylates. *J. Chromatogr. A.* 1997, 770, 51–57.

43. Podkrajsek B., Grgic, I., and Tursic J. Determination of sulfur oxides formed during the S(IV) oxidation in the presence of iron. *Chemosphere.* 2002, 49, 271–277.

44. Szabo, Z. et al. Analysis of nitrate ion in nettle (*Urtica dioica* L.) by ion-pair chromatographic method on a C30 stationary phase. *J. Agr. Food Chem.* 2006, 54, 4082–4086.

45. Varvaresou, A. et al. Development and validation of a reversed-phase ion-pair liquid chromatography method for the determination of magnesium ascorbyl phosphate and melatonin in cosmetic creams. *Anal. Chim. Acta* 2006, 573–574, 284–290.

46. Lin, S.H. et al. Peak crossover in high-performance liquid chromatography elution monitored using whole-column detection. *J. Chromatogr. A.* 2008, 1201 128–131.

12 Detection and Hyphenation

HPLC detectors deal with the time-dependent monitoring of column effluents. They signal, in the dimension of time, the positions of the analytes subjected to a chromatographic process. IPC detectors are those commonly used for classical HPLC separations but some special precautions should be taken to ensure compatibility between IPC conditions and detector mode of operation.

Current IPC detectors are on-stream monitors. HPLC detectors range from (1) non selective or universal (bulk property detectors such as the refractive index (RI) detector), characterized by limited sensitivity, (2) selective (discriminating solute property detectors such as UV-Vis detectors) to (3) specific (specific solute property detectors such as fluorescence detectors). Traditional detection techniques are based on analyte architecture that gives rise to high absorbance, fluorescence, or electrochemical activity. Mass spectrometry (MS) and evaporative light scattering detectors (ELSDs), can be considered universal types in their own right.

Detectors can be mass flow-sensitive (refractive index detectors, ELSDs) or concentration-sensitive (UV-Vis and fluorescence detectors). The former respond to the amount (mass) of analyte passing through the detector per unit of time (calculated by multiplying the eluent flow rate by the analyte concentration in the eluent), while the latter respond to analyte concentration. In mass flow-sensitive detectors, the response (signal amplitude) is proportional to the amount of sample component reaching the detector per unit of time.

Figure 12.1 clearly illustrates the popularity of diverse HPLC detectors among IPC practitioners. In 2008, more than 80% of all IPC applications were detected via mass spectrometry (MS) and/or ultraviolet-visible (UV-Vis) spectrophotometry. The role played by these two strategies is almost the same. Figure 12.1 shows advancements in HPLC detection, In recent decades, MS advanced from a niche, difficult-to-combine detection mode to a routine practice, as widespread as UV-Vis detection had been earlier. It is also striking to observe that RI detectors were not widely used, probably because their temperature sensitivity and incompatibility with gradient separations impaired their versatility. The RI detector was the first liquid chromatography (LC) type. It was very popular because it was considered a universal eluent bulk property detector because detection depended on the solute modifying the overall refractive index of the mobile phase; since its popularity is rapidly declining it will not be discussed further.

Detection is of crucial importance to achieve high sensitivity and prevent deterioration of peak sharpness obtained via efficient separation. It follows that strong interplay between detector performance and column performance is crucial. Increasing column efficiency, as indicated in Chapter 6, would be meaningless without controlling the extra-column source of band broadening. Miniaturization of the detector

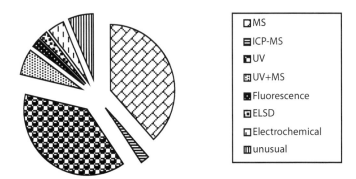

FIGURE 12.1 Diverse HPLC detectors used for IPC applications in 2008.

and conduit volume and redesign of detector geometry are prerequisites to properly handling small volume peaks and avoiding dispersion from dilution and Newtonian flow that could prevent adequate separation. Table 12.1 summarizes basic detector requirements and specifications.

12.1   UV-Vis DETECTORS

UV-Vis detectors respond to analytes that absorb light in the range from 190 to 750 nm. It is almost universal. If in the molecular architectures there double bonds or unshared electron pairs the molecule can absorb UV-Vis light. For reasons explained in Table 12.1, the column eluent passes through a very low volume cell (1 to 10 μl). Beer's law describes the relationship of the intensity of light transmitted through a cell (I_t), the intensity of light entering the cell (I_0), and the concentration of solute contained by it (c):

$$A = \log \frac{I_0}{I_t} = \varepsilon b c \tag{12.1}$$

where A is the absorbance, that is, the analytical signal, ε is the molar extinction coefficient of the solute (constant for the specific wavelength, temperature, and eluent composition), b is the path length of the cell, and c is the analyte concentration. The sensitivity of the detector will be directly proportional to εb. If b is increased to improve sensitivity, peak dispersion increases because the cell volume increases accordingly. A careful trade-off between sensitivity and peak broadening must be found. Careful selection of a UV-transparent organic modifier must be performed under gradient elution (water, methanol, acetonitrile, and tetrahydrofuran).

Fixed wavelength UV detectors are old-fashioned. Multi-wavelength detectors now make use of broad emission light sources (usually deuterium in the UV range, and tungsten for the visible range); a monochromator provides light of a specific wavelength and can be adjusted for specific detection purposes. Maximum sensitivity is obtained when the analyte absorbs light of a specific wavelength, while other solutes do not. The best sensitivity occurs when the selected wavelength of light corresponds

TABLE 12.1
Basic Detector Requirements

Detector Specification	Comment
Noise	Short time, high frequency fluctuation of baseline detector output from smooth, straight line that is not related to eluted solutes • Influenced by time-dependent pressure, temperature, eluent composition, and electrical signal fluctuation • Electronic filtering possible • Peak-to-peak measured by constructing parallel lines embracing maximum excursions of recorder trace over 15-min period
Drift	Long time (1 hr) deviation of baseline from horizontal line • Many detector-specific causes may be responsible of drift • Moderately harmless if due to detector warm-up after power on, drift of ambient temperature, or eluent flow rate
Sensitivity (S) For concentration-sensitive detectors: $S = AF/W$; for mass flow-sensitive detectors: $S = A/W$ where A = peak area, W = mass of analyte, F = eluent flow rate	Signal output (response) as peak area per unit concentration or unit mass flow of analyte • Corresponds to slope of calibration curve • Higher sensitivity lowers minimum amount that can be detected if noise is constant
Selectivity	Ability of detector to sense only components of interest, despite their possible co-elution with interferents • Higher selectivity lowers minimum amount that can be selected because it lowers signal noise
Minimum detectable amount $(MDA) = 3N/S$	Concentration or mass flow that gives detector signal equal to three times noise level • It can be lowered by increasing sensitivity and/or decreasing noise
Linear range (LR)	Range of concentrations or masses passing through detector over which detector signal is linear (constant sensitivity, usually within ±5%). • Minimum detectable amount and upper limit of linearity delimit LR • Numerically expressed by ratio of upper and lower limits
Dynamic range (DR)	Range of concentrations or masses passing through detector over which incremental change in concentration or mass produces incremental change in detector signal • Minimum detectable amount and upper limit of DR delimit e DR • DR > LR

(continued)

TABLE 12.1 (CONTINUED)
Basic Detector Requirements

Detector Specification	Comment
Time constant (τ)	Time required for the detector output to exponentially change from its initial value by the fraction $[1-\exp(-t/\tau)]$ (for t $=/\tau$) of the final value. • it hould be less than 50 milliseconds to avoid apparent dispersion of the eluted peak, but for special very fast separations, a lower value of 15 milliseconds may be necessary; time required for detector output to exponentially change from its initial value by fraction $[1-\exp(-t/\tau)]$ (for t $=\tau$) of final value, is called time constant
Sensor dimension	Apparent peak broadening if cell geometry not properly optimized or volume of detector cell is more than 1/10 of peak volume (may be calculated from efficiency and retention time) • Loss of chromatographic resolution • Tailing as result of stagnant zones

to the analyte absorption maximum. The dispersing element can be located before the sensor cell (classical dispersion detector) or after it (diode array detector [DAD]), as illustrated in Figure 12.2.

In the first case, if the whole spectrum of the column effluent must be recorded, a stop-flow procedure is required because the wavelength scan is somewhat slow. Moreover, if fluorescence is excited, the light falling on the photo cell that produces the signal will contain it, thereby prejudicing the linearity of the response. The DAD has gained widespread use because the inverse geometry of the dispersive element generally avoids the linearity problem. The dispersed light is arranged to fall on a linear diode array (hundreds of elements) and the absorption at very narrow ranges of wavelengths is continuously monitored at each diode, so that the complete spectrum of the column effluent can be continuously monitored, allowing the chromatographer to assess peak purity by comparing normalized spectra taken from both sides of the peak. In the case of a pure (spectrally homogeneous) peak, the ratio of the two spectra is unity. The resolution of the detector increases with the number of diodes in the array for the range of wavelengths covered. Excellent linearity range is a valuable feature of UV-Vis detection.

A three-dimensional spectrochromatogram can be also used for qualitative purposes and greatly increases detection selectivity. Two co-eluting analytes characterized by different absorption spectra may be quantified separately, using two carefully chosen monitoring wavelengths to avoid co-absorption. Absorption in the UV and visible range is still the standard detection mode even if the trend of detector popularity in recent years indicates UV-Vis will be surpassed by MS detection in the near future.

The next section will cover recent examples of the uses of UV-Vis detectors. The method has proven effective for identifying and dating fountain pen ink entries on documents [1], in a quality-by-design approach to impurity analysis of atomoxetine hydrochloride [2], for the simultaneous IPC determination of mycophenolic acid and its glucuronides in human plasma [3], and for the IPC quantification of zidovudine and its monophosphate [4]. Identification of the ellagic acid from oak leaves and sheep ruminal fluid was performed by analyzing retention time, UV absorption, and determination of purity [5].

Absorption in the UV and visible range is not universal because many analytes lack suitable strong chromophores. UV visualization of non-UV-absorbing compounds was often achieved via derivatization procedures aimed at modifying molecular architectures to make them suitable for UV-Vis light absorption. A vast assortment of derivatization strategies can be found among IPC applications. For example, metal ions were detected via a post-column derivatization with 4-(2-pyridylazo) resorcinol (PAR), and IPC quantitative results were in agreement with those obtained from atomic absorption spectroscopy [6]. Dithizone derivatives were used as chromogenic ligands for the quantitative determination of inorganic and organo-mercury analytes in aqueous matrices [7]. A very sensitive indirect detection of polythionates and thiosulfates relied on their catalytic effects of a post-column azide–iodine reaction [8]. Cysteine species in rat plasma were quantitatively analyzed via post-column ligand substitution by monitoring changes of absorbance at 500 nm [9]. Similarly, a post-column derivatization with 1,2-naphthoquinone-4-sulfonate proved useful for the determination of biogenic amines in wines [10]. The post-column use of a metallochromic reagent allowed the photometric detection of eluting lanthanides and actinides [11].

Pre-column derivatization is a valid alternative to post-column reaction; it is useful for improving detection and also modifying the ionic interactions of the analytes within the IPC system. The IPC determination of reduced and total mercaptamine in human urine with UV-Vis detection was based on pre-column derivatization with 2-chloro-1-methylquinolinium tetrafluoroborate [12]. The isocratic IPC isocratic assay of dihydroascorbic acid was possible after pre-column derivatization with 4,5-dimethyl-1,2-phenylenediamine, under mild conditions in the dark [13].

Detection problems solved using pre- or post-column derivatization involve time-consuming procedures and questionable quantification of the analytes because their reactions with the derivatization reagent may also not be quantitative. Indirect detection may represent a simpler solution to the lack of a suitable chromophore (or, generally speaking, the lack of analyte UV-Vis detectability). This strategy uses a detectable and retainable probe at a fixed concentration in the mobile phase. If the probe does not directly interact with the analyte when the analyte is injected, simple competition for adsorption sites gives rise to a detectable signal. If the IPR is the probe, a perturbation of the distribution equilibrium of the IPR between the mobile and the stationary phases arises upon injection of the analytes because of adsorption competition and ionic interactions. This perturbation produces a signal because the IPR is detectable; this allows the analyte to achieve indirect, non-specific detection.

The general theory of indirect detection was recently put forward [14] and its mechanism stimulated pioneering model makers to provide a realistic, practical and rational description of IPC [15,16]. For example, a UV-absorbing mobile phase containing potassium hydrogen phthalate and triethanolamine allowed the indirect detection of sulfur and nitrogen anions in atmospheric liquids separated on a cetylpyridinium-coated C18 column [17]. Crystal violet was used as the IPR in the separation of common inorganic anions detected down to 0.1 ppm at the absorption maximum of the dye [18].

12.2   FLUORESCENCE DETECTORS

When electromagnetic radiation of suitable wavelength is absorbed by a molecule, a transition to a higher electronic state occurs. Excited molecules can dissipate excess energy by collision with other molecules. When the electron returns to its ground state, a photon of a wavelength higher than that of the light absorbed is emitted. This phenomenon is photoluminescence. The emission can be immediate or delayed and the emitted substance is said to be fluorescent and phosphorescent, respectively.

The optical system of a fluorescence detector is arranged such that the source light falls on the eluent that passes through a small volume flow cell. Possible fluorescent light is sensed by a photoelectric cell, positioned normal to the direction of the incident light beam. At variance with a UV-Vis absorption detector in which the signal is superimposed on a strong background, the fluorescent signal is measured against very low background light (noise), that can also be further reduced by a filter via the removal of any stray scattered source light. This maximizes the signal-to-noise ratio and provides the lowest limit of detection (LOD) available in HPLC. Figure 12.2 compares the geometries of the variable wavelength UV-Vis, DAD, and fluorescence detectors. The fluorescence signal (I_f) can be obtained by the following relationship:

$$I_f = \varphi I_0 (1 - \exp(-\varepsilon b c)) \qquad (12.2)$$

where φ is the quantum yield and I, ε, b, and c have the same meanings as in Equation 12.1. Unfortunately the linearity of these detectors is limited, and since most substances do not naturally fluoresce, derivatization to accomplish fluorescence is a common practice. The simplest fluorescence detectors contain a single wavelength excitation source and a sensor that monitors fluorescent light of all wavelengths. They provide sensitive detection because the low pressure mercury lamp used as light source provides relatively high intensity UV light at 253.7 nm.

Light of this wavelength is particularly suitable to excite fluorescence in active molecules. Detectors with both selectable excitation and emission wavelengths are very complex and extremely versatile because they allow the optimization of the excitation and emission light wavelengths to obtain maximum sensitivity and also because excitation and fluorescence spectra can be obtained via the scan of the excitation and fluorescent lights respectively. This makes it possible to determine the best exciting wavelength for a given fluorescent light and the best fluorescent wavelength for a given exciting light, respectively. These spectra are also useful to confirm the identity an analyte. They require broad band excitation sources, usually deuterium lamps.

Sensitive and selective fluorescent IPC detection was recently used to analyze underivatized amino acids [19], albendazole marker residue in animal tissues [20], loratidine [21], early synthetic dyes [22], bitter orange alkaloids [23], and ochratoxin A in red wines [24].

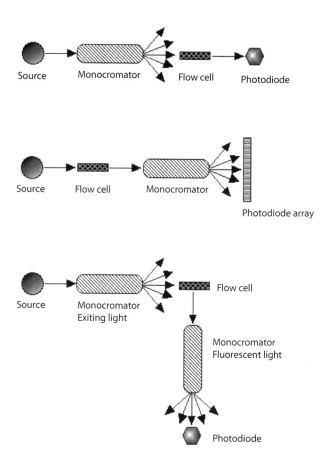

FIGURE 12.2 Comparison of UV-Vis (top), DAD (center), and (bottom) fluorescence detectors.

The lack of suitable chromophores for UV-Vis detection can be circumvented by derivatization, and the same strategy can be used to obtain the required fluorescence of an analyte if it does not naturally fluoresce. Many fluorophores were investigated and used. Determination of leukocyte DNA 6-thioguanine nucleotide levels relied on derivatization with chloroacetaldehyde [25]. A post-column derivatization with β-naphthoquinone-4-sulfonate allowed the detection of streptomycin and dihydrostreptomycin in foods [26]. A post-column derivatization with O-phthalaldehyde was used for the detection of biogenic amines and polyamines in vegetable products [27]; metal-cyanide complexes were detected after photodissociation and reaction of free cyanide by reaction with O-phthaldialdehyde and glycine [28]. Paralytic shellfish poisoning toxins as non-fluorescent tetrahydropurine compounds were post-column oxidized to fluorescent iminopurine derivatives [29]. Post-column nucleic acid intercalation of fluorescent dyes demonstrated more than a 1000-fold increase in sensitivity when compared with traditional nucleic acid UV detection [30]. Pre-column fluorescent labeling was also successfully explored for glycoprotein-derived oligosaccharides [31].

12.3   ELECTRICAL CONDUCTIVITY DETECTORS

This type of bulk property detector monitors the conductivity of the eluent. All ions from the analyte and from the buffer contribute to produce a signal. Detector response is linear over a wide range. Cell resistance is inversely proportional to electrolyte concentration. Since AC voltages must be used to avoid polarization of the sensing electrodes, the physical quantity measured is impedance, not resistance.

The sensor is a simple device consisting of two electrodes within a suitable flow cell of only a few microliters and an appropriate electronic circuit to produce the detector output. If the signal produced by buffer ions overwhelms the signal from analyte ions, an ion suppression column should be introduced between the column and the detector even if it involves band spreading that deteriorates the global resolving ability of the system. Membrane suppression technology is becoming more common because of its considerable enhancement of sensitivity and reduction of background conductivity [32]. Recent examples of analytes amenable to conductivity detectors include inorganic anions [33], aliphatic long chain quaternary ammonium compounds [34], and metallocyanide complexes of Fe(II), Ni(II), and Co(III) [35], among others.

12.4   ELECTROCHEMICAL DETECTORS

These detectors sense substances that undergo redox reactions at the surfaces of the electrodes. Aromatic rings with hydroxy and/or amino groups, secondary and tertiary amines, and aromatic nitrogen and thiol groups are all amenable to oxidation, while nitro and azo groups and aromatic or conjugated ketones easily undergo reduction; all are suitable for electrochemical detection. The detector signal arises from electron flow generated by the reaction. Mobile phases should be electrically conductive, free of metallic contaminants and oxygen to keep background current as low as possible. Normal phase systems that use apolar mobile phases are not suitable because they do not conduct. Adsorption of the oxidation or reduction products on the working electrode surface calls for frequent calibration and regular cleaning of the electrode system to ensure accurate quantitative analysis.

An electrochemical detector is destructive. It requires (1) a working electrode (where oxidation or reduction takes place), (2) an auxiliary electrode, and (3) a reference electrode (to regulate voltage and compensate for changes in background conductivity of the eluent). When an active substance flows into the electrochemical cell and a potential difference is applied between the working and reference electrodes, the electrolysis of the analyte yields a current (detector signal) that is a function of the applied potential. The three steps in the process are:

1. Diffusive mass transfer of the analyte (reagent) from bulk solution to electrode
2. Redox reaction at electrode surface
3. Mass transfer of reaction products from the electrode to the bulk solution

The first step is the slowest; the second is the fastest. The layer near the electrode is depleted of analyte. A concentration gradient between the electrode surface and the bulk of the solution is responsible for diffusion of the analyte into the depleted zone at a rate proportional to its eluent concentration. Since the slowest step determines the overall reaction rate, the current (i) generated at the electrode surface is diffusion controlled.

The relationship between the diffusion controlled current and the potential is a voltammogram characteristic of each substance that may be used also for qualitative purposes. The diffusion controlled current always depends linearly on solute concentration and increases with increased diffusivity, the number of electrons involved in the redox reaction, and eluent flow rate. It also depends on the cell and electrode geometries. When the surface area of the electrode is raised, both the signal (i) and the noise increase. A reduction in sensor size produces a significant increase in signal-to-noise ratio with a consequent lowering of the LOD [36]. Miniaturization is useful to control peak dispersion in the sensor and involves also higher flow rates that contribute to increasing the signal even if the sensor is very flow sensitive and a constant flow rate is required.

The three most common modes of operation of electrochemical detection are amperometric, coulometric, and potentiometric. An amperometric detector is an electrochemical cell that produces a signal proportional to the analyte concentration; usually the percentage of the analyte that undergoes the redox reaction is very low, about 5%.

The coulometric detector is a particular type of amperometric detector in which the percentage of electrolyzed analyte is almost 100%. The area of the peak on a current versus time plot (charge flow during electrolysis) is related to the analyte concentration via Faraday's law. This technique cannot be used with very active compounds because the analyte products would pollute the electrode surface if the reaction proceeded to completion (exhausting all the reactant).

Poppe [37] described many possible geometric forms of the electrodes. The choice for electrode material is usually a carbon paste based on its mechanical ruggedness and long-term stability.

Vitreous or "glassy" carbon is an excellent electrode material due to its compatibility with organic solvents. The porous graphitic carbon electrode characterized by a very high specific surface area was frequently used in electrochemical detectors because it was mechanically robust and permeable to the mobile phase. Flow-through electrodes were devised and commercialized. They proved valuable because their surface areas greatly exceed areas needed for quantitative electrochemical reactions also under coulometric conditions. Flow-through electrodes may be up to 95% contaminated before they fail to function. More important, if cleaning of the electrodes is required (usually after 1 to 3 years of constant use), the contamination can be eliminated rapidly by flushing with nitric acid.

Figure 12.3 shows an electrode unit system. Each unit has a central porous carbon electrode between a reference electrode and an auxiliary electrode. Because the porous electrode is permeable, many electrode units can be connected in series forming an array without creating a strong pressure drop across the array. Commercially available sensors consist of up to 16 units and a sensor system consisting of 80 electrodes in the space of a few millimeters has been assembled [38]. Gradually increasing potential is applied consecutively to the electrode units. Analytes flow through the array until each arrives at the unit that has the required potential to make it undergo a redox reaction.

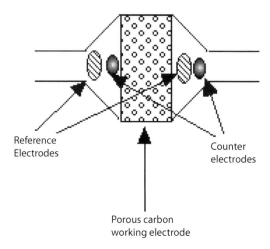

Reference
Electrodes

Counter
electrodes

Porous carbon
working electrode

FIGURE 12.3   Electrode unit of electrochemical detector.

Coulometric array detection offers an additional resolution dimension because the analyte is completely reacted over three contiguous electrode units that cover a potential range within the half potential of the voltammogram typical of that analyte. Further units no longer sense the analyte. Conversely, in amperometric detection, only a small part of the analyte reacts and the balance continues to be sensed in subsequent units. Coulometric array detectors may produce three-dimensional graphs. Since the ratios of signals from the three adjacent electrode units where the solute reacts will be different for diverse analytes (although they may be sensed at the same three electrodes) this method is capable of verifying the identity of a putative solute.

The coulometric array detection mimics the diode array detector; if two peaks are eluted together, they can be electrochemically and spectrophotometrically resolved and quantified. While a diode array detector relies on the different UV-Vis spectra of various compounds and characterizes analytes on the basis of their retention times and spectroscopic features, the coulometric array detector takes advantage of the variability of the voltammograms for diverse analytes and typifies them based on retention time and reaction potential.

Iodide in urine [39] and catecholamines [40,41] are example of analytes recently detected electrochemically and studied under IPC conditions. Pulsed amperometric detection on a gold electrode was used to detect etimicin [42] and gentamicin [43] in commercial samples, thus avoiding tedious pre-column derivatization. Heterocyclic aromatic amines in soup cubes [44] were determined by a coulometric electrode array detector, and the coulometric detection of a quinone-bearing drug candidate [45] allowed the study of electrochemical properties.

Potentiometric monitoring was also explored. Solid-state ion-selective electrodes were tested as sensing elements to detect diverse amines. The incorporation of molecular recognition host molecules in the membrane improved the detection limits and the sensitivity was found to be better than that obtained with indirect UV detection. This potentiometric detection system is interesting because it avoids the need for derivatization and post-column reactions [46].

12.5   EVAPORATIVE LIGHT SCATTERING DETECTORS

The evaporative light scattering detector (ELSD) [47] is based on the ability of fine particulate matter of a solute to scatter light. To obtain suitable analyte particles, the column effluent is nebulized by an inert gas in the nebulizer and aerosol droplets are allowed to evaporate in the drift tube. Droplet size is related to mobile phase properties (surface tension, density, and viscosity). Usually, high solvent-to-gas flow ratio provides the best sensitivity because it produces the largest droplet diameters.

However, the gas flow rate must be optimized because it is inversely proportional to droplet size. Larger droplets scatter more light, thereby increasing the analytical signal and sensitivity, but they are also less prone to evaporate, thereby increasing baseline noise. Since the limit of detection is related to the signal-to-noise ratio, the chromatographer must seek the best compromise. Droplet size must be carefully controlled because it establishes the size of the dried solute particles that in turn determines the extent of the linear response. It is important to use the lowest possible evaporative temperature to allow solute crystal formation, to avoid evaporation of solute and destruction of thermo-sensitive solute.

The fine particulate matter of the solute (or droplets for liquid analytes) suspended in the nebulizer gas pass through a laser beam and the Rayleigh scattered light, proportional to the size and number of particles, is sensed perpendicularly (or at an angle) to the incident light beam by a photodiode. The minimum detectable quantity is 10 to 20 ng of solute. The detector responds to all solutes that are less volatile than the mobile phase [48] and the signal is proportional to the mass flow of the analytes. This technique is currently used as a quasi-universal detector that eliminates the need for derivatization of non-UV-absorbing analytes. Actually, ELSD was used as a surrogate MS detector during method development, albeit with lower sensitivity, since volatile mobile phases are prerequisite for the evaporation of the mobile phase prior to both the light scattering step and MS detection. Volatile (formic, acetic, and trifluoroacetic) acids or bases (triethylamine, ammonia) and their salts can be used as IPRs. They were used under IPC conditions to analyze underivatized amino acids [49], pharmaceuticals samples [50,51], and oligo-ι-carrageenans [52] and room temperature ionic liquids employed as enzymatic reaction media [53]. In comparison to MS, ELSD requires less investment and lower operating expenses and offers easy operation and less maintenance.

12.6   UNUSUAL DETECTORS

Among unusual detection techniques, we highlight photoluminescence following electron transfer and chemiluminescence. The former was used to detect biogenic monoamines in rat brain striatum microdialysates separated under IPC conditions from interfering acids [54]; the latter proved very sensitive for the determination of dithiocarbamate fungicides by IPC [55] and pipecolic acid that, like other imino acids, showed strong chemiluminescence upon mixing with tris(2,2′-bipyridine) ruthenium(III) [56].

A whole-column detection system fabricated with parts from a typical A4 size optical scanner was used to monitor peak crossover that occurs when a solute moves slower than another one, then during gradient elution, migrates faster and reaches the column outlet earlier. A 0.3-mm spatial resolution and a 3.6 ms temporal resolution were obtained and proved adequate for directly monitoring solute retention behavior in a liquid chromatography column under IPC conditions [57]. On line radiochemical detection was also explored [58].

12.7   HYPHENATION TECHNIQUES

12.7.1   IPC-MS

The MS detection technique is based on the ionization and fragmentation of an analyte. The mass analyzer separates the molecular ion and fragments according to their mass-to-charge ratios. Usually the detector output represents the total ion current produced by all fragments and molecular ions. Of course, this technique can be considered universal because all molecules can be sensed. However, if a particular ion that epitomizes a certain analyte (or class of analytes) is specifically sensed (via selected ion monitoring operation) while all other ions are disregarded, the technique becomes very specific.

Joining IPC to MS techniques probably provides the most valuable detection mode because it complements the chromatographic separation of complex mixtures with structural information and unequivocal identification of analyte along with good sensitivity [59–65]. For example, it unequivocally identified paralytic shellfish poisoning toxins [29] and heterocyclic aromatic amines [66] and distinguished positional isomers [67]. The presence of ionic modifiers in the mobile phase under IPC conditions epitomizes IPC-MS technique and makes it differ from HPLC-MS. Table 12.2 shows the contrasting requirements of IPC and MS.

TABLE 12.2
Contrasting Requirements of IPC and MS

IPC	MS
Liquid phase process	Vacuum process
1 ml/min eluent flow ≈ 500ml/min of gas at flow STP	Accepts only 10 ml/min gas flow
Inorganic buffers and IPRs	Volatile mobile phase components
25 to 50°C	200 to 350°C
All analytes	Volatile analytes

From the analysis of the specifications in Table 12.2, IPC – MS hyphenation requires: (1) specific optimization of mobile phase composition, selecting only volatile components, (2) devising strategies to minimize eluent flow, and (3) an interface to make IPC effluent amenable to MS detection. The goals of the interface are (1) separation of the analyte from the bulk eluent, (2) ion evaporation for ionic species or ionization of non-ionic solutes, and (3) fragmentation and quantitative transfer of analyte fragments to the mass analyzer.

An ideal interface should not cause extra-column peak broadening. Historical interfaces include the moving belt and the thermospray. Common interfaces are electrospray ionization (ESI) and atmospheric pressure chemical ionization (APCI). Several special interfaces include the particle beam—a pioneering technique that is still used because it is the only one that can provide electron ionization mass spectra. Others are continuous flow fast atom bombardment (CF-FAB), atmospheric pressure photon ionization (APPI), and matrix-assisted laser desorption ionization (MALDI). The two most common interfaces, ESI and APCI, were discovered in the late 1980s and involve an atmospheric pressure ionization (API) step. Both are soft ionization techniques that cause little or no fragmentation; hence a fingerprint for qualitative identification is usually not apparent.

In ESI, the analyte is introduced to the interface at a flow rate of a few microliters per minute and passes through the capillary of an electrospray needle. The high potential difference with respect to the counter electrode (3 to 5kV) on the sampling cone forces the spraying of charged droplets from the needle at ambient pressure. Since the surface charge of the droplets has the same polarity as the charge on the needle, they are attracted toward the oppositely charged source sampling cone. Concurrently, solvent evaporation occurs and the droplet shrinks until a "coulombic explosion" arising from electrostatic repulsion occurs, thereby producing singly or multiply charged analyte molecules.

The analyte reaches the mass analyzer (10^{-5} Torr) through a capillary (1 Torr) and a skimmer, led by electrostatic lenses. Multiply charged analytes allow analysis of high molecular weight components such as biopolymers because the mass-to-charge ratio is decreased and falls within the measurement range of the MS. Since ESI is a soft ionization technique, it provides the molecular weight (MW) of the analyte because very little fragmentation is produced. Heated ESI yielded better results in metabolomics [68]. ESI is clearly the interface of choice for analytes that are already ionic in the eluent, such as those typically analyzed under IPC conditions.

Ionic modifiers in the IPC eluent can provide a charge for neutral molecules so that they become detectable via the ESI interface. Conversely, counter ions reduce the charge state of an oppositely charged analyte or even convert it to the opposite polarity. If they are polycharged, for example, ESI$^+$ and ESI$^-$ modes were comparatively evaluated for the detection of nucleotides that are negatively charged. It was straightforward to use the ESI$^-$ mode that detects the $[M - H]^-$ ion (with low levels of sodium and potassium adducts present), but ESI$^+$ was a viable alternative because the volatile N,N-dimethylhexylamine IPR yields ion-pairs with the nucleotides. The most abundant relevant ion was the adduct between the compound and the IPR agent, along with the $[M + H]^+$ ion [69]. The eluent pH also influences the best (positive or negative) ESI mode [71].

APCI involves chemical ionization at atmospheric pressure. The general source set-up is similar to that for ESI except that ionization occurs separately in APCI. The analyte is instantaneously desolvated; the effluent solution is introduced into a pneumatic nebulizer and a heated quartz tube helps solvent evaporation before interacting with the corona discharge that creates ions. The corona discharge electrode (6 kV) produces primary nitrogen ions by electron ionization. They collide with the vaporized solvent molecules to form secondary reactant gas ions. The eluent solvents should produce gas phase acid–base reactions with the analyte. In the positive mode, the analyte should have a proton affinity higher than that of the solvent to take its proton; in the negative mode, the opposite holds; the proton should be transferred from the analyte to the solvent. Repeated collisions of solvent ions with the analyte result in the formation of analyte ions that enter the pumping and focusing stages before mass analysis, similar to ESI. Table 12.3 compares the most common interfaces used for IPC-MS and suggests positive and negative features of both API techniques.

The mass analyzer separates the molecular ions and fragments according to their mass-to-charge ratios. Common mass analyzers are the quadrupole, ion trap (IT), time-of-flight (TOF), magnetic sector, and ion cyclotron resonance (ICR) analyzers.

TABLE 12.3

Positive (+) and Negative (–) Features of Most Common Interfaces Used for IPC-MS Combination

ESI	APCI
+ Info on MW	+ Info on MW
+ Polycharged ions, good for high MW analytes	Not good for high MW analytes
+ Good also for non-volatile and thermally labile analytes	+ Good only for volatile and thermally stable analytes
– Scarce fragmentation	– No fragmentation
– Requires capillary column (eluent flow rate <5µl/min) or split devices	+ Accepts eluent flow rates up to 1 ml/min
– Ion suppression in presence of high eluent ionic strength	–Eluent solvents should be able to give gas phase acid–base reactions with analyte
– Not good for high water percentages in eluent, requires volatile mobile phase components	– Requires volatile mobile phase components
Preferred for anionic (negative mode) and at ionic (positive mode) analytes or very polar analytes	Preferred for polar analytes (apolar analytes better analyzed via GC-MS)

ESI = Electrospray Ionization.
APCI = Atmospheric Pressure Chemical Ionization

The mass analyzers used most frequently for IPC-MS are the quadrupole, TOF, and IT because they are inexpensive and provide good dynamic ranges. Additional positive attributes of IT are extended mass range and improved resolution compared to those features of the quadrupole. Both ESI and APCI produce little or no fragmentation; this in turn reveals minimal molecular architecture details. This problem may be resolved through the use of tandem mass spectrometric techniques such as MS/MS or MS^n in which a collision gas rips apart each already analyzed precursor ion (collision-induced dissociation [CID]). The resulting product ions are further analyzed via mass spectrometry. The analysis of a single "product" ion is known as selected reaction monitoring (SRM). The technique of monitoring two or three specific fragment or "product" ions is known as multiple reaction monitoring (MRM). Along with selected ion recording, it is possible to find all parent ions of a given daughter ion or all daughters ion of a parent. MRM allows the recognition of all solute that loses a given fragment. For example, it proved valuable for Se speciation of Brazil nuts because the transition of 198–102 is specific for Se-containing molecules since it corresponds to the loss of a Se fragments [72].

Many instrumental set-ups and geometries have been explored. In triple quadrupole mass spectrometry, the first quadrupole selects the parent ion of interest, the second works as a collision cell to fragment the parent ion, and the third isolates the proper product ion. A hybrid type is the quadrupole time-of-flight (Q-TOF) instrument.

ESI is the most common interface since IPC and MS were coupled initially. By 2008, most applications IPC–MS used the ESI interface [58,68–82] because analytes amenable to IPC are usually already ionic in the column effluent that enters the interface. Examples of APCI-MS applications [83,84] include two-fold use of both interfaces [85] they gave similar results in the determination of polyunsatured fatty acid monoepoxides [86]. For determining mono- and di-sulfonated azo dyes, ESI proved to give the best performance in terms of sensitivity and reproducibility [83]. Joining negative APCI-MS and ESI-MS unambiguously identified several acidic oxidation products of 2,4,6-trinitrotoluene in ammunition, wastewater, and soil extracts [61].

Recent examples of IPC-MS combination confirm that this multifaceted strategy is particularly expedient in critical fields such as pharmaceutical and medical analyses [73–76,79,80]. This hyphenated technique can handle complex samples and was used to measure DNA inter-strand cross-linking efficiency [81], increase the efficiency of forensic DNA fingerprinting [82], and analyze underivatized amino acids in geological samples [77]. The utility of IPC-MS was recently demonstrated for targeted metabolomics by triple quadrupole mass spectrometry operating in MRM mode. High mass resolution full scan mass spectrometry is preferred for bridging targeted and untargeted metabolomics [68].

IPC-MS combined methods must be optimized with respect to separation and compatibility with online detection involving the constraints detailed in Table 12.2 regarding the composition and volatility of the mobile phase. The major concern of chromatographers who deal with this combined technique is the reduced signal caused by source pollution of non-volatile IPRs. Moreover, the efficiency of droplet development, which in turn affects the number of charged ions that ultimately reach the detector, may be prejudiced due to the formation of ion-pairs between the analyte and IPR [87]. Severe ion suppression can impair the sensitivity of detection [88] and some unique strategies were devised to eliminate these difficulties.

First, volatile IPRs rather than traditional IPRs were usually selected during method development. Although non-volatile tetraalkylammonium salts generally provide the best chromatographic selectivity, volatile di- and trialkylammonium acetates or formats offer good trade-offs between retention and selectivity on one hand and electrospray signal on the other [70,83,86,89–93]. Ammonium acetate was also used [79,80,83,94] but reportedly caused similar signal suppression as di and tri-alkylammonium acetates with poorer separation selectivity [95]. It is interesting that ion suppression effects are related to instrument geometries in the order of Z spray < orthogonal spray < linear spray [95]. The most favorable chromatographic conditions depend strongly on the mass spectrometer used, and more sensitive instruments are usually less robust [96].

Perfluorinated anions, a class of volatile IPRs, are earning increasing credibility for use in IPC-ESI-MS [70,73,74,76,77,87,89,97–99]. Perfluorinated IPRs and trifluoroacetic (TFA) in particular are becoming the most frequently used IPRs for IPC of proteins and peptides. They simultaneously control eluent pH, improve resolution and peak shape by reducing silanol interactions, and enhance retention via ion-pairing. Gustavsson and co-workers proved that the chromatographic performances of TFA, heptafluorobutanoic acid (HFBA), and perfluoroheptanoic acid were as good as those of classical sulfonate IPRs. Moreover, the performance of the ESI interface was not impaired after 24 hours of continuous infusion. HFBA produced better sensitivity than TFA, but while analyte retention increased with increasing HFBA concentration, the ESI signal was suppressed [87]. Unfortunately, the elevated surface tension of these reagents impairs efficient spray formation. To keep the signal suppression as low as possible and mitigate the adverse effect, a post-column addition of another volatile such as propionic acid was successfully explored [99]. Kwon and Moini used atmospheric microwave-induced plasma ionization MS, which relies on gas-phase reactions rather than solution ionization, to overcome this problem when using nonafluoro pentanoic acid (NFPA) as a mobile-phase additive [100].

Since volatile IPRs may fail to provide adequate retention and separation [61], other approaches for combining IPC and MS were investigated. The favorable retention and selectivity characteristics of tetrabutylammonium (TBA) may be exploited by using volatile counter ions such as acetate [101] or bromide [102]. Another approach is the online removal of non-volatile ions by ion exchange so that the combination may be accomplished without contamination of the interface [84,103]. The use of a nanosplitting device with high splitting ratios was also suggested to reduce signal suppression [104].

It is possible to perform an IPC separation with an IPR-free mobile phase, at least under certain conditions. The IPR can be added only to the sample solution to be injected into the chromatographic system. This strategy, explained in

Chapter 11, is valuable because it dramatically reduces operation costs because no IPR is needed in the eluent and IPC-MS is straightforward. The absence of IPR in the eluent minimizes ionization suppression in the MS source, thus increasing the sensitivity of the method [60,105–107].

A further advantage of combining IPC and MS is that isotope-labeled internal standards may be used for accurate quantification of analytes. No recovery check is needed for analytes because the labeled analogues of the analytes that serve as internal standards are added to the samples before the extraction step. This allows sample mixtures to be analyzed without pretreatment because the labeled internal standards share the fate of the unlabeled analyte during sample processing without differences in extraction efficiency between the labeled standards and unlabeled analytes. This method achieved excellent precision and improved linearity of calibration lines despite interference from sample matrices in the quantitative analysis of important secondary intracellular metabolites in a complex biological sample solution from cultured cells. The technique is a valuable strategy for metabolomics [75].

12.7.2 IPC-ICP AND OTHER UNUSUAL HYPHENATIONS

A combination of IPC and inductively coupled plasma (ICP) MS was extensively explored for the speciation of phosphorus, arsenic, selenium, cadmium, mercury, and chromium compounds [108–118] because it provides specific and sensitive element detection. Selenium IPC speciation was joined to atomic fluorescent spectrometry via an interface in which all selenium species were reduced by thiourea before conventional hydride generation [119]. Coupling IPC separation of monomethyl and mercuric Hg in biotic samples by formation of their thiourea complexes with cold vapor generation and atomic fluorescence detection was successfully validated [120]. The coupling of IPC with atomic absorption spectrometry was also used for online speciation of Cr(III) and Cr(VI) [121] and arsenic compounds employing hydride generation [122].

In the analysis of nucleotide-activated sugars, structural information was gained via IPC-ESI-MS and IPC-NMR. The NMR spectrometer was equipped with a flow cell; a stop-flow valve allowed precise parking of the peak of interest to perform the NMR scan [123].

Online coupling of surface-enhanced resonance Raman spectroscopy and IPC proved valuable for the identification of basic dyes [124]. Circular dichroism spectroscopy is an extraordinary technique for selective detection of compounds possessing optically active adsorption bands and was successfully coupled to IPC of steroids [125]. Table 12.4. summarizes the most important features and parameters to be compared when selecting a detection mode for an IPC application.

TABLE 12.4
Characteristics of Common Detectors (Data from manufacturers' literature)

	RI	UV-Vis	Fluorescence	Electrochemical	Conductivity	ELSD	MS
Range of application	Universal	Selective	Very selective for fluorescent compounds	Very selective for electroactive compounds	Very selective for ionic compounds	Universal for all solutes less volatile than eluent	Universal and very selective; depends on operation mode
Structural data	No	Limited	Limited	Limited	No	No	Complete identification of compounds
Sensitive to:	Concentration	Concentration	Concentration	Concentration	Concentration	Mass flow	Mass flow
Sensitivity	Low	High	Very High	High	High	Medium	Medium
Minimum detectable quantity (g)	10^{-7}	10^{-10}	10^{-12}	10^{-11}	10^{-11}	10^{-9}	10^{-11}
Linear range (upper to lower limit ratio)	10^4	10^5	10^4	10^6	10^4	10^3	10^3
Gradient compatibility	No	Yes	Yes	No	No	Yes	Yes
Temperature sensitivity	High	Low	Low	High	High	Low	Low
Cost	Low	Low	Low	Low	Low	Medium	High

REFERENCES

1. Wang, X.F. et al. Identification and dating of the fountain pen ink entries on documents by ion-pairing high-performance liquid chromatography. *Forensic Sci. Int.* 2008, 180, 43–49.
2. Gavin, P.F. and Olsen, B.A. Quality by design approach to impurity method development for atomoxetine hydrochloride (LY139603). *J. Pharm. Biomed. Anal.* 2008, 46, 431–441.
3. Mino, Y. et al. Simultaneous determination of mycophenolic acid and its glucuronides in human plasma using isocratic ion-pair high-performance liquid chromatography. *J. Pharm. Biomed.* 2008, 46, 603–608.
4. Lefebvre, I. et al. Quantification of zidovudine and its monophosphate in cell extracts by on-line solid-phase extraction coupled to liquid chromatography. *J. Chromatogr. B.* 2007, 858, 2–7.
5. del Moral, P.G. et al. Determination of ellagic acid in oak leaves and in sheep ruminal fluid by ion-pair RP-HPLC. *J. Chromatogr. B.* 2007, 855, 276–279.
6. Pobozy, E. et al. Flow-injection sample preconcentration for ion-pair chromatography of trace metals in waters. *Water Res.* 2003, 37, 2019–2026.
7. Shaw M.J., Jones P., and Haddad, P.R. Dithizone derivatives as sensitive water-soluble chromogenic reagents for ion chromatographic determination of inorganic and organo-mercury in aqueous matrices. *Analyst* 2003, 128, 1209–1212.
8. Miura, Y. and Watanabe, M. Ion-pair chromatography of polythionates and thiosulfate with detection based on their catalytic effects on the post-column azide–iodine reaction. *J. Chromatogr. A.* 2001, 920, 163–171.
9. Harada, D. et al. Determination of reduced, protein-unbound, and total concentrations of N-acetyl-L-cysteine and L-cysteine in rat plasma by post-column ligand substitution high performance liquid chromatography. *Anal. Biochem.* 2001, 290, 251–259.
10. Hlabangana, L., Hernandez-Cassou, S., and Saurina J. Determination of biogenic amines in wines by ion-pair liquid chromatography and post-column derivatization with 1,2-naphthoquinone-4-sulphonate. *J. Chromatogr. A.* 2006, 1130, 130–136.
11. Jaison, P.G., Raut, N.M., and Aggarwal, S.K. Direct determination of lanthanides in simulated irradiated thoria fuels using reversed-phase high-performance liquid chromatography. *J. Chromatogr. A.* 2006, 1122, 47–53.
12. Kusémierek, K. and Bald, E. Measurement of reduced and total mercaptamine in urine using liquid chromatography with ultraviolet detection. *Biomed. Chromatogr.* 2008, 22, 441–445.
13. Gioia, M.G. et al. Development and validation of a liquid chromatographic method for the determination of ascorbic acid, dehydroascorbic acid and acetaminophen in pharmaceuticals. *J. Pharm. Biomed.* 2008, 48, 331–339.
14. Forssén, P. and Fornstedt, T. General theory of indirect detection in chromatography. *J. Chromatogr. A.* 2006, 1126, 268–275.
15. Bidlingmeyer, B.A. et al. Retention mechanism for reversed-phase ion-pair liquid chromatography. *J. Chromatogr.* 1979, 186, 419–434.
16. Bidlingmeyer, B.A. Separation of ionic compounds by reversed-phase liquid chromatography: update of ion-pairing techniques. *J. Chromatogr. Sci.* 1980, 18, 525–539.
17. Zuo, Y. and Chen, H. Simultaneous determination of sulfite, sulfate, and hydroxymethane-sulfonate in atmospheric waters by ion-pair HPLC technique. *Talanta* 2003, 59, 875–881.
18. Tonelli, D., Zappoli, S., and Ballarin, B. Dye-coated stationary-phases: retention model for anions in ion-interaction chromatography. *Chromatographia* 1998, 48, 190–196.
19. Tripp, J.A., McCullagh, J.S.O., and Hedges, R.E.M. Preparative separation of underivatized amino acids for compound-specific stable isotope analysis and radiocarbon dating of hydrolyzed bone collagen. *J. Sep. Sci.* 2006, 29, 41–48.

20. Fletouris, D.J. et al. Highly sensitive ion-pair liquid chromatographic determination of albendazole marker residue in animal tissues. *J. Agr. Food Chem.* 2005, 53, 893–898.

21. Sora, D.I. et al. Validated ion-pair liquid chromatography and fluorescence detection method for assessing the variability of loratadine metabolism occurring in bioequivalence studies. *Biomed. Chromatogr.* 2007, 21, 1023–1029.

22. van Bommel, M.R. et al. High-performance liquid chromatography and non-destructive three-dimensional fluorescence analysis of early synthetic dyes. *J. Chromatogr. A.* 2007, 1157, 260–272.

23. Putzbach, K. et al. Determination of bitter orange alkaloids in dietary supplement standard reference materials by liquid chromatography with ultraviolet absorbance and fluorescence detection. *J. Chromatogr. A.* 2007, 1156, 304–311.

24. Yu, J.C.C. and Lai, E.P.C. Determination of ochratoxin A in red wines by multiple pulsed elutions from molecularly imprinted polypyrrole. *Food Chem.* 2007, 105, 301–310.

25. Olesen, K.M. et al. Determination of leukocyte DNA 6-thioguanine nucleotide levels by high-performance liquid chromatography with fluorescence detection. *J. Chromatogr. B.* 2008, 864, 149–155.

26. Viñas, P., Balsalobre, N., and Hernández-Córdoba, M. Liquid chromatography on an amide stationary phase with post-column derivatization and fluorimetric detection for the determination of streptomycin and dihydrostreptomycin in foods. *Talanta* 2007, 72, 808–812.

27. Lavizzari, T. et al. Improved method for the determination of biogenic amines and polyamines in vegetable products by ion-pair high-performance liquid chromatography. *J. Chromatogr. A.* 2006, 1129, 67–72.

28. Miralles, E. et al. Determination of metal–cyanide complexes by ion-interaction chromatography with fluorimetric detection. *Anal. Chim. Acta* 2000, 403, 197–204.

29. Krock, B., Seguel, C.G., and Cembella, A.D.Toxin profile of *Alexandrium catenella* from the Chilean coast as determined by liquid chromatography with fluorescence detection and liquid chromatography coupled with tandem mass spectrometry. *Harm. Algae* 2007, 6, 734–744.

30. Dickman, M.J. Post-column nucleic acid intercalation for the fluorescent detection of nucleic acids using ion-pair reverse phase high-performance liquid chromatography. *Anal. Biochem.* 2007, 360, 282–287.

31. Yoshinaka, Y., Ueda, Y., and Suzuki, S. Ion-pair chromatographic separation of glycoprotein-derived oligosaccharides as their 8-aminopyrene-1,3,6-trisulfonic acid derivatives. *J. Chromatogr. A.* 2007, 1143, 83–87.

32. Haddad, P.R. Ion chromatography retrospective. *Anal. Chem.* 2001, 73, 266A–273A.

33. Hatsis, P. and Lucy, C.A. Ultra-fast HPLC separation of common anions using a monolithic stationary phase. *Analyst* 2002, 127, 451–454.

34. Giovannelli, D. and Abballe, F. Aliphatic long chain quaternary ammonium compound analysis by ion-pair chromatography coupled with suppressed conductivity and UV detection in lysing reagents for blood cell analysers. *J. Chromatogr. A.* 2005, 1085, 86–90.

35. Souza e Silva, R. et al. Separation and determination of metallocyanide complexes of Fe(II), Ni(II) and Co(III) by ion-interaction chromatography with membrane suppressed conductivity detection applied to analysis of oil refinery streams (sour water). *J. Chromatogr. A.* 2006, 1127, 200–206.

36. Weber, S.G. and Purdy, W.C. Electrochemical detectors in liquid chromatography: a short review of detector design. *Ind. Eng. Chem. Prod. Res. Dev.* 1981, 20, 593–598.

37. Poppe, H. Electrochemical detectors. *In Instrumentation for High-Performance Liquid Chromatography,* Huber, J.F.K., Ed. Elsevier: Amsterdam, 1978, pp. 131–149.

38. Atsushi, A., Matsue T., and Uchida, I. Multichannel electrochemical detection with microelectrode array in flowing streams. *Anal. Chem.* 1992, 64, 44–49.

39. Below, H. and Kahlert, H. Determination of iodide in urine by ion-pair chromatography with electrochemical detection. *Fresenius J. Anal. Chem.* 2001, 371, 431–436.

40. Baranyi, M. et al. Chromatographic analysis of dopamine metabolism in a Parkinsonian model. *J. Chromatogr. A.* 2006, 1120, 13–20.

41. Lee, M. et al. Selective solid-phase extraction of catecholamines by chemically modified polymeric adsorbents with crown ether, *J. Chromatogr. A.* 2007, 1160, 340–344.

42. Xi, L., Wu, G., and Zhu, Y. Analysis of etimicin sulfate by liquid chromatography with pulsed amperometric detection. *J. Chromatogr. A.* 2006, 1115, 202–207.

43. Manyanga, V. et al. Improved liquid chromatographic method with pulsed electro-chemical detection for analysis of gentamicin, *J. Chromatogr. A.* 2008, 1189, 347–354.

44. Krach, C. and Sontag, G. Determination of some heterocyclic aromatic amines in soup cubes by ion-pair chromatography with coulometric electrode array detection. *Anal. Chim. Acta* 2000, 417, 77–83.

45. Mancini, F. et al. Monolithic stationary phase coupled with coulometric detection: development of an ion-pair HPLC method for the analysis of quinone-bearing compounds. *J. Sep. Sci.* 2007, 30, 2935–2942.

46. Poles, I. and Nagels, L.J. Potentiometric detection of amines in ion chromatography using macrocycle-based liquid membrane electrodes. *Anal. Chim. Acta* 2001, 440, 89–98.

47. Megoulas, N.C. and Koupparis, M.A. Twenty years of evaporative light scattering detection. *Crit. Rev. Anal. Chem.* 2005, 35, 301–316.

48. Petritis, K., Elfakir, C., and Dreux, M. Comparative study of commercial liquid chromatographic detectors for the analysis of underivatized amino acids. *J. Chromatogr. A.* 2002, 961, 9–21.

49. Petritis, K. et al. Validation of an ion-interaction chromatography analysis of underivatized amino acids in commercial preparation using evaporative light scattering detection. *Chromatographia* 2004, 60, 293–298.

50. Sarri, A.K., Megoulas, N.C., and Koupparis M.A. Development of a novel liquid chromatography–evaporative light scattering detection method for bacitracins and applications to quality control of pharmaceuticals. *Anal. Chim. Acta* 2006, 573–574, 250–257.

51. Xie, Z., Jiang, Y., and Zhang, D. Simple analysis of four bisphosphonates simultaneously by reverse phase liquid chromatography using n-amylamine as volatile ion-pairing agent. *J. Chromatogr. A.* 2006, 1104, 173–178.

52. Antonopoulos, A. et al. Characterisation of iota-carrageenans oligosaccharides with high-performance liquid chromatography coupled with evaporative light scattering detection. *J. Chromatogr. A.* 2004, 1059, 83–87.

53. Lue, B.M., Guo, Z., and Xu, X. High-performance liquid chromatography analysis methods developed for quantifying enzymatic esterification of flavonoids in ionic liquids. *J. Chromatogr. A.* 2008, 1198–1199 107–114.

54. Jung, M.C. et al. Simultaneous determination of biogenic monoamines in rat brain dialysates using capillary high-performance liquid chromatography with photoluminescence following electron transfer. *Anal. Chem.* 2006, 78, 1755–1760.

55. Nakazawa, H. et al. Determination of dithiocarbamate fungicides by reversed-phase ion-pair liquid chromatography with chemiluminescence detection. *J. Liq. Chromatogr. Rel. Technol.* 2004, 27, 705–713.

56. Kodamatani, H. et al. Highly sensitive and simple method for measurement of pipecolic acid using reverse-phase ion-pair high performance liquid chromatography with tris(2,2′-bipyridine) ruthenium(III) chemiluminescence detection. *J. Chromatogr. A.* 2007, 1140, 88–94.

57. Lin, S.H. et al. Peak crossover in high-performance liquid chromatography elution monitored using whole-column detection. *J. Chromatogr. A.* 2008, 1201, 128–131.

58. Koitka, M. et al. Determination of rat serum esterase activities by an HPLC method using s-acetylthiocholine iodide and p-nitrophenyl acetate. *Anal. Biochem.* 2008, 381, 113–122.

59. Holzl, G. et al. Analysis of biological and synthetic ribonucleic acids by liquid chromatography-mass spectrometry using monolithic capillary columns. *Anal. Chem.* 2005, 77, 673–680.

60. Gao, S. et al. Evaluation of volatile ion-pair reagents for the liquid chromatography–mass spectrometry analysis of polar compounds and its application to the determination of methadone in human plasma. *J. Pharm. Biomed. Anal.* 2006, 40, 679–688.

61. Schmidt, T.C., Buetehorn, U., and Steinbach, K. HPLC-MS investigations of acidic contaminants in ammunition wastes using volatile ion-pairing reagents (VIP-LC-MS). *Anal. Bioanal. Chem.* 2004, 378, 926–931.

62. Chen, H.C. and Ding, W.H. Hot-water and solid-phase extraction of fluorescent whitening agents in paper materials and infant clothes followed by unequivocal determination with ion-pair chromatography-tandem mass spectrometry. *J. Chromatogr. A.* 2006, 1108, 202–207.

63. Liu, Y.Z. et al. Studies on the degradation of blue gel pen dyes by ion-pairing high performance liquid chromatography and electrospray tandem mass spectrometry. *J. Chromatogr. A.* 2006, 1125, 95–103.

64. Oberacher, H. et al. Some guidelines for the analysis of genomic DNA by PCR-LC-ESI-MS. *J. Am. Soc. Mass Spectr.* 2006, 17, 124–129.

65. Jin, F. et al. Determination of diallyldimethylammonium chloride in drinking water by reversed-phase ion-pair chromatography–electrospray ionization mass spectrometry. *J. Chromatogr. A.* 2006, 1101, 222–225.

66. Calbiani, F. et al. Validation of an ion-pair liquid chromatography–electrospray-tandem mass spectrometry method for the determination of heterocyclic aromatic amines in meat-based infant foods. *Food Addit. Contam.* 2007, 24, 833–841.

67. Papousková, B. et al. Mass spectrometric study of selected precursors and degradation products of chemical warfare agents *J. Mass Spectrom.* 2007, 42, 1550–1561.

68. Lu, W., Bennett, B.D., and Rabinowitz, J.D. Analytical strategies for LC-MS-based targeted metabolomics. *J. Chromatogr. B.* 2008, 871, 236–242.

69. Cordell, R.L. et al. Quantitative profiling of nucleotides and related phosphate-containing metabolites in cultured mammalian cells by liquid chromatography tandem electrospray mass spectrometry. *J. Chromatogr. B.* 2008, 871, 115–124.

70. Warnkea, M.M. et al. The Evaluation and Comparison of Trigonal and Linear Tricationic Ion-Pairing Reagents for the Detection of Anions in Positive Mode ESI-MS *J. Am. Soc. Mass Spectrom.* 2009, 20, 529–538.

71. Toll, H. et al. Separation, detection, and identification of peptides by ion-pair reversed-phase high-performance liquid chromatography-electrospray ionization mass spectrometry at high and low pH. *J. Chromatogr. A.* 2005, 1079, 274–286.

72. Dumont, E. et al. Speciation of Se in *Bertholletia excelsa* (Brazil nut): a hard nut to crack? *Food Chem.* 2006, 95, 684–692.

73. Eckstein, J.A. et al. Analysis of glutamine, glutamate, pyroglutamate, and GABA in cerebrospinal fluid using ion-pairing HPLC with positive electrospray LC/MS/MS. *J. Neurosci. Meth.* 2008, 171, 190–196.

74. Prokai, L. et al. Measurement of acetylcholine in rat brain microdialysates by LC isotope dilution tandem MS. *Chromatographia* 2008, 68, 5101–5105.

75. Seifar, R.M. et al. Quantitative analysis of metabolites in complex biological samples using ion-pair reversed-phase liquid chromatography-isotope dilution tandem mass spectrometry *J. Chromatogr. A.* 2008, 1187, 103–110.

76. Häkkinen, M.R. et al. Quantitative determination of underivatized polyamines by using isotope dilution RP-LC-ESI-MS/MS. *J. Pharm. Biomed. Anal.* 2008, 48, 414–421.

77. Liu, D.L., Beegle, L.W., and Kanik, I. Analysis of underivatized amino acids in geological samples using ion-pairing liquid chromatography and electrospray tandem mass spectrometry. *Astrobiology* 2008, 8, 229–241.

78. Ghassempour, A. et al. Monitoring of N-nitrosodiethanolamine in cosmetic products by ion-pair complex liquid chromatography and identification with negative ion electrospray ionization mass spectrometry. *J. Chromatogr. A.* 2008, 1185, 43–48.

79. Liu, Y.Q. et al. Quantitative determination of erythromycylamine in human plasma by liquid chromatography-mass spectrometry and its application in a bioequivalence study of dirithromycin. *J. Chromatogr. B.* 2008, 864, 1–8.

80. Bharathi, V.D. et al. LC-MS-MS assay for simultaneous quantification of fexofenadine and pseudoephedrine in human plasma. *Chromatographia* 2008, 67, 461–466.

81. Narayanaswamy, M. et al. An assay combining high-performance liquid chromatography and mass spectrometry to measure DNA interstrand cross-linking efficiency in oligonucleotides of varying sequences. *Anal. Biochem.* 2008, 374, 173–181.

82. Oberacher, H. et al. Increased forensic efficiency of DNA fingerprints through simultaneous resolution of length and nucleotide variability by high-performance mass spectrometry. *Human Mutat.* 2008, 29, 427–432.

83. Rafols, C. and Barcelo, D. Determination of mono- and disulphonated azo dyes by liquid chromatography–atmospheric pressure ionization mass spectrometry. *J. Chromatogr. A.* 1997, 777, 177–192.

84. Socher, G. et al. Analysis of sulfonated compounds by reversed-phase ion-pair chromatography-mass spectrometry with on-line removal of non-volatile tetrabutyl ammonium ion-pairing agents. *Chromatographia* 2001, 54, 65–70.

85. Castro, R., Moyano, E., and Galceran M.T. On-line ion-pair solid-phase extraction-liquid chromatography–mass spectrometry for the analysis of quaternary ammonium herbicides. *J. Chromatogr. A.* 2000, 869, 441–449.

86. Fer, M. et al. Determination of polyunsatured fatty acid monoepoxides by high performance liquid chromatography–mass spectrometry. *J. Chromatogr. A.* 2006, 1115, 1–7.

87. Gustavsson, S.A. et al. Studies of signal suppression in liquid chromatography–electrospray ionization mass spectrometry using volatile ion-pairing reagents. *J. Chromatogr. A.* 2001, 937, 41–47.

88. Takino, M., Daishima, S., and Yamaguchi, K. Determination of diquat and paraquat in water by liquid chromatography–electrospray mass spectrometry using volatile ion-pairing reagents. *Anal. Sci.* 2000, 16, 707–711.

89. Ariffin, M.M. and Anderson, R.A. LC/MS/MS analysis of quaternary ammonium drugs and herbicides in whole blood. *J. Chromatogr. B.* 2006, 842, 91–97.

90. Vas, G. et al. Study of transaldolase deficiency in urine samples by capillary LC-MS/MS. *J. Mass Spectrom.* 2006, 41, 463–469.

91. Coulier, L. et al. Simultaneous quantitative analysis of metabolites using ion-pair liquid chromatography electrospray ionization mass spectrometry. *Anal. Chem.* 2006, 78, 6573–6582.

92. Shen, X. et al. Investigation of copper–azoamacrocyclic complexes by high-performance liquid chromatography. *Biomed. Chromatogr.* 2006, 20, 37–47.

93. Vanerkova, D. et al. Analysis of electrochemical degradation products of sulphonated azo dyes using high-performance liquid chromatography/tandem mass spectrometry. *Rapid Commun. Mass Spectrom.* 2006, 20, 2807–2815.

94. Baiocchi, C. et al. Characterization of methyl orange and its photocatalytic degradation products by HPLC/UV-Vis diode array and atmospheric pressure ionization quadrupole ion trap mass spectrometry. *Int. J. Mass Spectrom.* 2002, 214, 247–256.

95. Holcapek, M. et al. Effects of ion-pairing reagents on the electrospray signal suppression of sulphonated dyes and intermediates. *J. Mass Spectrom.* 2004, 39, 43–50.

96. De Person, M., Chaimbault, P., and Elfakir, C. Analysis of native amino acids by liquid chromatography/electrospray ionization mass spectrometry: comparative study between two sources and interfaces. *J. Mass Spectrom.* 2008, 43, 204–215.

97. Aramendia, M.A. et al. Determination of diquat and paraquat in olive oil by ion-pair liquid chromatography–electrospray ionization mass spectrometry. *Food Chem.* 2006, 97, 181–188.

98. Piraud, M. et al. Ion-pairing reversed-phase liquid chromatography/electrospray ionization mass spectrometric analysis of 76 underivatized amino acids of biological interest: a new tool for the diagnosis of inherited disorders of amino acid metabolism. *Rapid Commun. Mass Spectrom.* 2005, 19, 1587–1602.

99. Häkkinen, M.R. et al. Analysis of underivatized polyamines by reversed phase liquid chromatography with electrospray tandem mass spectrometry. *J. Pharm. Biomed. Anal.* 2007, 45, 625–634.

100. Kwon, J. and Moini, M. Analysis of underivatized amino acid mixtures using high performance liquid chromatography/dual oscillating nebulizer atmospheric pressure microwave induced plasma ionization-mass spectrometry. *J. Am. Soc. Mass Spectrom.* 2001, 12, 117–122.

101. Gibson, C.R. et al. Electrospray ionization mass spectrometry coupled to reversed-phase ion-pair high performance liquid chromatography for quantitation of sodium borocaptate and application to pharmacokinetic analysis. *Anal. Chem.* 2002, 74, 2394–2399.

102. Witters, E. et al. Ion-pair liquid chromatography–electrospray mass spectrometry for the analysis of cyclic nucleotides. *J. Chromatogr. B.* 1997, 694, 55–63.

103. Henriksen, J., Roepstorff, P., and Ringborg, L.H. Ion-pairing reversed-phased chromatography/mass spectrometry of heparin. *Carbohydr. Res.* 2006, 341, 382–387.

104. Gangl, E.T. et al. Reduction of signal suppression effects in ESI-MS using a nanosplitting device. *Anal. Chem.* 2001, 73, 5635–5644.

105. Marín, J.M. et al. An ion-pairing liquid chromatography/tandem mass spectrometric method for the determination of ethephon residues in vegetables. *Rapid Commun. Mass Spectrom.* 2006, 20, 419–426.

106. Ghassempour, A. et al. Monitoring of N-nitrosodiethanolamine in cosmetic products by ion-pair complex liquid chromatography and identification with negative ion electrospray ionization mass spectrometry. *J. Chromatogr. A.* 2008, 1185, 43–48.

107. Ding, W., Hill, J.J., and Kelly, J. Selective enrichment of glycopeptides from glycoprotein digests using ion-pairing normal phase liquid chromatography. *Anal. Chem.* 2007, 79, 8891–8899.

108. Helfrich, A. and Bettmer, J. Determination of phytic acid and its degradation products by ion-pair chromatography (IPC) coupled to inductively coupled plasma-sector field-mass spectrometry (ICP-SF-MS). *J. Anal. Atom. Spectrom.* 2004, 19, 1330–1334.

109. Kapolna, E. et al. Selenium speciation studies in Se-enriched chives (*Allium schoenoprasum*) by HPLC-ICP–MS. *Food Chem.* 2007, 101, 1398–1406.

110. Vonderheide, A.P. et al. Investigation of selenium-containing root exudates of *Brassica juncea* using HPLC-ICP-MS and ESI-qTOF-MS. *Analyst* 2006, 131, 33–40.

111. Loreti, V. et al. Biosynthesis of Cd-bound phytochelatins by *Phaeodactylum tricornutum* and their speciation by size-exclusion chromatography and ion-pair chromatography coupled to ICP-MS. *Anal. Bioanal. Chem.* 2005, 383, 398–403.

112. Richardson, D.D., Sadi, B.B.M., and Caruso, J.A. Reversed phase ion-pairing HPLC-ICP-MS for analysis of organophosphorus chemical warfare agent degradation products. *J. Anal. Atom. Spectrom.* 2006, 21, 396–403.

113. Grotti, M. et al. Arsenobetaine is a significant arsenical constituent of the red Antarctic alga *Phyllophora antarctica*. *Environ. Chem.* 2008, 5, 171–175.

114. Yathavakilla, S.K.V. and Caruso, J.A. Study of Se-Hg antagonism in *Glycine max* (soybean) roots by size exclusion and reversed phase HPLC-ICPMS. *Anal. Bioanal. Chem.* 2007, 389, 715–723.

115. Vallant, B., Kadnar, R., and Goessler, W. Development of a new HPLC method for the determination of inorganic and methylmercury in biological samples with ICP-MS detection. *J. Anal. Atom. Spectrom.* 2007, 22, 322–325.

116. Zheng, J. Shibata, Y. Tanaka, A. Study of the stability of selenium compounds in human urine and determination by mixed ion-pair reversed-phase chromatography with ICP-MS detection. *Anal. Bioanal Chem.* 2002, 374, 348–353.

117. Pan, F., Tyson, J.F., and Uden, P.C. Simultaneous speciation of arsenic and selenium in human urine by high-performance liquid chromatography inductively coupled plasma mass spectrometry. *J. Anal. Atom. Spectrom.* 2007, 22, 931–937.

118. Kápolna, E. et al. Selenium speciation studies in Se-enriched chives (*Allium schoenoprasum*) by HPLC-ICP-MS. *Food Chem.* 2007, 101, 1398–1406.

119. Qiu, J. et al. On-line pre-reduction of Se(VI) by thiourea for selenium speciation by hydride generation. *Spectrochim. Acta B.* 2006, 61, 803–809.

120. Shade, C.W. Automated simultaneous analysis of monomethyl and mercuric Hg in biotic samples by Hg-thiourea complex liquid chromatography following acidic thiourea leaching. *Environ. Sci. Technol.* 2008, 42, 6604–6610.

121. Sarica, D.Y., Turker, A.R., and Erol, E. On-line speciation and determination of Cr(III) and Cr(VI) in drinking and waste water samples by reversed-phase high performance liquid chromatography coupled with atomic absorption spectrometry. *J. Sep. Sci.* 2006, 29, 1600–1606.

122. Ko, F.H., Chen, S.L., and Yang, M.H. Evaluation of the gas–liquid separation efficiency of a tubular membrane and determination of arsenic species in groundwater by liquid chromatography coupled with hydride generation atomic absorption spectrometry. *J. Anal. Atom. Spectrom.* 1997, 12, 589–595.

123. Ramm, M. et al. Rapid analysis of nucleotide-activated sugars by high-performance liquid chromatography coupled with diode-array detection, electrospray ionization mass spectrometry and nuclear magnetic resonance. *J. Chromatogr. A.* 2004, 1034, 139–148.

124. Seifar, R.M. et al. At-line coupling of surface-enhanced resonance Raman spectroscopy and reversed-phase ion-pair chromatography. *Anal. Commun.* 1999, 36, 273–276.

125. Gergely, A. et al. Evaluation of CD detection in an HPLC system for analysis of DHEA and related steroids. *Anal. Bioanal. Chem.* 2006, 384, 1506–1510.

13 Examples of Applications

Separation scientists have been hard pressed to keep pace with the explosion of analysis numbers brought about by massive advances in the food, life science, medicine, pharmacology, toxicology, environmental, agricultural and biological sectors and to address fundamental security and safety issues. Further challenges surround the ever-increasing complexity of samples that demand versatile separation strategies.

Figure 13.1 illustrates the major fields of application of IPC in recent years. This chapter aims to impose order on the complex welter of IPC separations; it gives a brief overview and outlook covering recent applications of IPC in diverse analytical fields. For the sake of brevity, only the most significant recent publications on analytical applications are cited; we made no attempt to include an exhaustive survey of the literature. Information about other examples can be found elsewhere [1].

13.1 INORGANIC AND ORGANOMETALLIC SPECIES

The number and broad coverage of new IPC applications relating to inorganic samples indicates the versatility of this technique that earlier dealt with organic ions. Since Ion Chromatography (IC) is less efficient than HPLC in certain areas, many investigators focus on IPC to separate inorganic ions and organometallic species, usually analyzed by IC via ion-exchanging processes. Speciation—the determination of various forms of the same element—is a demanding task for analytical chemists. Many techniques cannot discriminate the different molecular architectures that represent the same element or cannot distinguish its diverse oxidation states. IPC may smooth the progress of speciation because it is separative in its own right.

One of the most exploited procedures is the complexation of metal ions. When a ligand bears a chromophore, a favorable side effect of this procedure is that sensitivity with conventional detection UV-Vis can be improved [2]. Speciation of Cr(III) and Cr(VI) in waters relied on the pre-complexation of Cr(III) with EDTA and IPC-ICP (inductively coupled plasma)-MS separation of the negatively charged chromium species in less than 2 minutes using a tetrabutylammonium IPR [3].

IPC-ICP-MS was extensively explored for the speciation of phosphorus, arsenic, selenium, cadmium, mercury, and chromium compounds [4–17] because it provides specific and sensitive element detection. Selenium IPC speciation took also advantage of coupling with atomic fluorescent spectrometry via an interface in which all selenium species are reduced by thiourea before conventional hydride generation [15]. IPC separation of monomethyl and mercuric Hg in biotic samples by formation of their thiourea complexes, coupled to cold vapor generation and atomic fluorescence detection, was successfully validated [18]. The coupling of IPC with atomic absorption spectrometry was also used for online speciation of arsenic compounds employing hydride generation [17]. In the analytical speciation of chromium using in

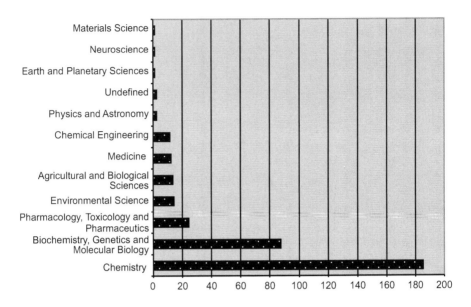

FIGURE 13.1   Fields of application of IPC from 2006 through 2008.

vitro cultures of chromate-resistant filamentous fungi for chromate bioremediation, IPC and anion exchange chromatography yielded consistent results, indicating that speciation of chromium in biological systems should not be limited to its two most common oxidation states [19].

Complex formation is useful for metal speciation and also for the separation of diverse metal ions. Among a variety of complexing reagents [20–22] cyanide is probably the most important. IPC separation of metal ions as metallocyanide complexes with a suitable cationic IPR is a reliable technique [23]. Complexation of trace level lanthanides with α-hydroxy isobutyric acid and separation under IPC condition shortened analysis time from days to minutes [24]. Flow injection was successfully coupled to IPC to simplify batch precomplexation; detection limits were at sub-microgram per liter levels [2].

Anion analysis continues to be significant in many scientific and technical fields. A general and sensitive method for detecting many singly charged anions (nitrate, thiocyanate, perchlorate, perfluorooctanoic acid, halogenated acetic acids, and other inorganic and organic anions) relied on the use of a dicationic reagent to form a complex that holds an overall positive charge for analysis by MS. The method provides the lowest reported detection limits for a variety of anions, especially chaotropes, and can speciate related species [25].

Separations of common inorganic anions were carried out on three different surfactant-coated columns converted into anion exchange stationary phases by equilibration with cationic surfactants [26]. Several polarizable anions were analyzed under IPC conditions via sulfonium and phosphonium IPRs that showed distinct selectivity toward the anions based on the chaotropic or kosmotropic attitudes of both the anion and the IPR [27].

A recent IPC determination of the main contaminants (bromide, iodide, sulfite, thiosulfate, thiocyanate, iron(III)-EDTA chelate, free EDTA, hydroquinone, and phenidone) in spent photographic solutions was optimized by varying the concentrations of tetrabutylammonium IPR, phosphate, acetonitrile, pH, and ionic strength. The separation of five inorganic anions, two metal chelates and two neutral compounds was successfully accomplished in about 30 minutes [28].

13.2   FOOD ANALYSIS

Many classes of analytes relevant for the food industry and food quality control are amenable to IPC. Vitamins were determined under IPC condition in a variety of matrices; for example, the IPC determination of total vitamin C in a broad range of fortified food products yielded results in agreement with the official titrimetric method [29].

IPC-MS/MS was used to quantify heterocyclic aromatic amines in meat-based infant foods [30]. The separation of biogenic amines was chemometrically optimized when they were determined in wines [31]; a sensitive and selective method to determine 12 biogenic amines regardless of the characteristics of the vegetal food matrix was successfully validated [32]. Determination of soybean proteins in commercial products was performed by fast IPC using an elution gradient and acetic acid as the IPR [33].

A very sensitive and specific IPC-ESI-MS method for determining cyclamate was used to monitor the content of the artificial sweetener in foods [34]. Similarly, IPC proved valuable for controlling the separation of complex mixtures of betacyanin and decarboxylated betacyanin food colorants via gradient elution [35].

Food safety concerns were also addressed via IPC. Many veterinary drugs were controlled in a broad range of food matrices. For example, a specific IPC strategy was developed for routine determination of the marker residues of oxytetracycline in edible animal tissues [36]. Dihydrostreptomycin is an aminoglycoside antibiotic used in veterinary medicine in combination with benzylpenicillin to treat bacterial infections in cattle, pigs and sheep; its IPC quantitative determination in bovine tissues and milk was developed and optimized [37,38].

Furosine, a marker of the Maillard reaction product, is a valuable indicator of food protein quality. It is a marker for thermal treatment in foodstuffs and is directly related to the loss of lysine availability. IPC was employed to determine furosine content in beverages based on soy milk and cow milk supplemented with soy isoflavones [39]. Furosine was also analyzed in 60 commercial breakfast cereals to assess their protein nutritional values. The higher the protein content in the formulation, the higher the furosine levels [40]. A simple IPC technique that uses 1-octanesulfonic acid as the IPR allowed the selective determination of histamine levels in fermented food [41].

13.3   LIFE SCIENCE AND MEDICINE

IPC serves as a powerful tool in many fields of life science. Nucleotides, nucleosides, and related compounds play important roles as structural units of nucleic acids, as co-enzymes in biochemical pathways, and as sources of chemical energy. Nucleotides, nucleosides, oligonucleotides, oligodeoxynucleotides, and nucleobases [42–54], DNA fragments [55], and DNA and RNA [56] were recently analyzed via IPC. The potential of ionic liquids as ionic modifiers for the analysis of nucleic compounds was recently demonstrated [57].

Since alterations in global DNA methylation are implicated in various pathobiological processes, a gradient IPC-ESI-MS/MS method with a volatile IPR was used to determine cytosine and 5-methylcytosine in DNA; quantification relied on stable isotope dilution [58]. Muscular dystrophies caused by various mutations in the dystrophin gene are amenable to easier prenatal diagnosis via a multiplex polymerase chain reaction (PCR)/IPC assay [59]. Some guidelines for the analysis of genomic DNA by IPC-ESI-MS can be found in Reference 60.

Proteomics also rely on IPC. While the array of strategies available for quantitative proteomics continues to expand, the number of attendant bioinformatic software tools also continues to grow. High-throughput peptide and protein identification technologies based on mass spectra involve IPC-MS/MS in combination with database searching algorithms. In order to exploit the information gained from the time domains of the separation processes, prediction models based on a new kernel function were devised, thereby providing a retention time filter. This approach required only very small training sets and improved the fractions of correctly identified peptides significantly [61]. Similarly, a sequence-specific retention calculator for peptide retention time prediction under IPC conditions was proposed. [62]. The crucial roles of peptides in several fields, for example, pharmacology, clinical diagnosis, and biomedical research are now widely accepted.

Glycopeptides from a peptide mixture were analyzed via normal phase IPC, using inorganic monovalent ions as IPRs (added only to the sample and not to the mobile phase) to increase the hydrophobicity differences among peptides and glycopeptides [63]. IPC separation of peptides usually relies on the use of perfluorinated carboxylic acids [64] and small chaotropes [65,66]. It can be performed also on a preparative scale for the purification of peptides from natural sources [67]. Amino acids and related compounds [68–72] were similarly analyzed via IPC-MS.

IPC proved valuable in metabolomics that represents the most ambitious addition to the applied genomics and proteomics toolbox. Metabolomics require the detection of all metabolites in a biological system, consequently a comprehensive analytical platform must be developed to fill this challenging need. Recent reports have demonstrated that IPC with an amine IPR is often a useful method for separating a broad range of negatively charged metabolites including nucleotides, sugar phosphates, and carboxylic acids. A two-dimensional-(size exclusion-IPC) ICP-MS and ESI MS/MS allowed the characterization of a selenocysteine-containing metabolome in selenium-rich yeast [73]. A rapid, specific, and reliable IPC assay of fermentation acid metabolites of lactic acid bacteria was developed [74]. The IPC-MS combination has proven useful in metabolomics. Its applicability to targeted metabolomics via triple quadrupole mass spectrometry operating in MRM mode was recently demonstrated although high mass resolution full scan mass spectrometry is necessary to bridge targeted and untargeted metabolomics [75]. For example a rapid, sensitive, and selective IPC-ESI-MS/MS method was developed for quantitative analysis of free intracellular metabolites in cell cultures. The isotope dilution technique (isotope labeled internal standards) avoided complex extraction [76]. A highly selective and sensitive IPC-ESI-MS/MS method was developed to identify and quantify intracellular metabolites involved in central carbon metabolism (including glycolysis, pentose phosphate pathways, and the tricarboxylic acid cycle) [77].

13.4 PHARMACEUTICAL, TOXICOLOGICAL, AND CLINICAL ANALYSIS

The pharmaceutical industry now faces compelling demands to improve the productivity of analyses and shorten drug discovery and development time cycles. An impressive number of drug candidates generated by modern synthesis and combinatorial libraries requires screening of many samples. Some examples of the usefulness of the IPC strategy will be discussed below.

Separations of antibiotics were studied thoroughly. An IPC method using perfluorocarboxylic acids as IPRs and pulsed electrochemical detection was devised to analyze gentamicin and performed better than the official method prescribed by the European Pharmacopoeia [78].

An IPC-ESI-MS/MS method allowed the simultaneous determination of neomycin and bacitracin in human and rabbit sera [79] and the analysis of aminoglycoside antibiotics in human plasma [80]. IPC was recently validated for the estimation of bulk and formulated gatifloxacin [81]. The IPC determination of norfloxacin in diverse matrices worked as a stability indicating method [82]. A C12 stationary phase with embedded polar group successfully achieved IPC baseline tetracycline separation simply by using a phosphate as the IPR [83]. A practical IPC method for the quality control of fosfomycin calcium and its related substances was recently suggested [84].

Bisphosphonates (bone resorption inhibitor drugs) were subject to many investigations. They can chelate to metal surfaces, producing chromatographic peak tailing. Different tetraalkylammonium salts, commonly selected as IPRs in separation of bisphosphonates [85], were replaced by volatile organic amines when ELSD was used [86].

The effects of chaotropic mobile phase additives, including ionic liquids [87], on retention behavior of β-blockers were studied and revealed that they influence selectivity and efficiency [88]. The biological activities of heparin and heparan sulfate make them significant pharmaceutical targets. They bind numerous proteins including growth factors and cytokines, and mediate various biological processes. Ion-pair UPLC (IP-UPLC), coupled with electrospray time-of-flight (TOF) mass spectrometry played a crucial role in developing fast and sensitive analytical techniques for the characterization and compositional analysis of these compounds at the disaccharide level [89].

IPC is also effective for pharmacology related to brain disease. It was used to analyze picolinic acid and related compounds [90], neurotoxins associated with paralytic shellfish poisoning [91], a drug candidate for treating Alzheimer's disease [92], and nicergoline, clinically used for improving brain metabolism [93]. Quaternary ammonium anticholinergics were determined in whole blood and the matrix effect was taken into account [94].

An IPC procedure assessed recoveries of urinary catecholamines during an innovative sample clean-up [95] and was optimized to avoid interferences by anti-TB drugs [96]. Adrenergic amines were determined in a variety of bitter orange-containing dietary supplements marketed as appetite suppressants; a sodium dodecyl sulfate IPR and fluorescent detection were used [97]. Similarly, two classes of compounds, alkaloids and flavonoid analytes, from citrus herbs used to treat obesity were analyzed by IPC-ESI-MS [98]. Simultaneous quantification of fexofenadine and pseudoephedrine in human plasma was also optimized via IPC [99]. The retention activities of selected alkaloids were studied using chaotropic IPRs whose behaviors agreed with their ranks in the Hofmeister series [100].

Amphetamines were widely investigated via IPC. A fully automated method using online solid phase extraction and IPC-ESI-MS/MS was developed to analyze amphetamine, methamphetamine, 3,4-methylenedioxyamphetamine, 3,4-methylene-dioxyethylamphetamine, and 3,4-methylenedioxymethamphetamine in urine samples. Trifluoroacetic acid served as the IPR to effectively minimize silanophilic interactions. The new method was validated against a literature GC/MS method. Good agreement was obtained from analysis of urine samples from drug users, with detection limits ranging from 1 to 3 ng/ml for each analyte [101]. A direct IPC determination of p-hydroxymethamphetamine glucuronide in human urine was also reported [102].

An IPC-ESI-MS/MS method using volatile perfluorinated carboxylic acids as IPRs added directly to the sample solution (not incorporated into the mobile phase) was developed for the quantitative assay of methadone in human plasma. This cost-effective strategy enhanced the efficiency of separation and minimized ion suppression because no IPR was present in the eluent [103]. Table 13.1 lists other potential uses of IPC in the pharmaceutical and clinical analysis fields.

13.5   ENVIRONMENTAL ANALYSIS

The 20th century was characterized by dramatic and beneficial advances in science and technology for the benefit of mankind. Unfortunately, the extraordinary progress achieved evolved over many years and imposed long-term environmental costs that future generations will inherit. Early in the new century, it became clear that any further progress in science and technology must include protection of the environment. The release of environmentally damaging compounds continues to be a problem of the chemical industry and "green chemistry" is the focus of interest of many research groups. Meanwhile, analytical chemists continue to develop experimental techniques and applications to monitor substances that pose concerns from an environmental view. IPC proved to be a valuable tool to analyze pesticides in various matrices [94,104] and phenoxyacetic herbicides and their main metabolites in soil samples [105].

An IPC-MS/MS spectrometric method to determine ethephon residues in vegetables included an IPR-free mobile phase; the IPR was added directly to the sample solution to minimize ionization suppression in the MS source and increase the sensitivity of the method [106].

Ion-pair extraction and IPC were combined to analyze phosphoric acid mono- and diesters originating from the microbial hydrolysis of flame retardants. Even tertiary treatment did not ensure complete removal of the studied compounds detected in municipal wastewater [107]. Chlorophenols extracted from water samples as anionic chlorophenolates were studied by IPC because the anionic forms of these analytes provide better UV ultraviolet absorption than uncharged chlorophenol based on their auxochromic effects. IPC conditions yielded adequate retention of the charged analytes and good sensitivity [108].

TABLE 13.1
Additional Examples of IPC Applications

Analyte	Matrix	IPR	Detection	Ref.
Erythromycylamine	Plasma	Ammonium acetate	MS	131
Mycophenolic acid	Human plasma	Tetra-n-butylammonium bromide containing 5 mM ammonium acetate	UV	132
Amodiaquine and its active metabolite	Human blood	Pentafluoropropionic acid	MS/MS	133
Zodovudine (AZT) and its monophosphate	Cell extracts	Triethylammonium acetate	UV	134
Loratadine (descarboethoxyloratadine)	Human plasma	Sodium dodecyl sulfate	Fluorescence	135
Acetylcholine	Rat brain microdialysis samples	Trifluoroacetic acid	MS/MS	136
Degraded products of nerve agents	Aqueous extract of soil sample	Tri-n-butyl amine	ESI-MS	137
Imatinib (tyrosine kinase inhibitor) and its main metabolite	Human plasma	1-Octane sulfonic acid	UV	138.
Rosuvastatin	Human plasma	Formic acid	MS/MS	139
Malonyl-CoA and seven other short-chain acyl-CoAs	Rat and mouse tissues	Dimethylbutyl amine	TOFMS	140
Diastereomers of novel HCV serine Protease inhibitor	Monkey plasma	Perfluoropentanoic acid	MS/MS	141
d-Penicillamine and tiopronin	Urine samples during treatment	Trichloroacetic acid	Derivatization, UV	142
Simvastatin (SV),	–	–	UV	143
Ethacridine lactate	Human urine	Sodium dodecyl sulfonate	UV	144
Chondroitin sulfate	Raw materials and dietary supplements		UV	145
Adefovir nucleotide analogue	Cell culture	Tetrabutylammonium and ammonium phosphate	MS/MS	146
Cytarabine	Mouse plasma		ESI-MS/MS	147
Pergolide (ergot alkaloid derivative used to treat Parkinson's disease)	Aqueous solutions and tablets	Sodium octanosulfonate	UV	148

In view of the widespread industrial use of numerous chelating agents, an IPC method was optimized for the selective analysis of 14 agents in wastewater to monitor their environmental fate [109]. The formation of the Fe(III) complexes of common chelating agents followed by IPC with a volatile ion-pairing agent and ESI-MS/MS allowed the direct injection of most aqueous environmental samples (wastewater, surface water, and drinking water) without preparatory enrichment. The method indicated that EDTA was an omnipresent contaminant in partially closed water cycles [110]. Organophosphorus chemical warfare degradation products were amenable to hyphenated IPC-ICP-MS/MS [111].

Several analytes were monitored in the effluents of potable water treatment plants. Poly-diallyldimethylammonium chloride is a water-soluble cationic polymer widely used as a coagulant in potable water treatment; it was analyzed by IPC-ESI-MS without sample preparation [112]. A method for the trace determination of fluorescent whitening agents was developed, and analytes were identified and quantified by IPC-ESI-MS/MS used with a volatile IPR. The method was applied to environmental water samples, showing the occurrence of five fluorescent whitening agents in wastewater treatment plant effluent samples and river water [113]. IPC methods have also analyzed azo dyes and monitored their degradation. Azo dyes are synthetic and they are highly persistent in the environment. They resist direct aerobic bacterial degradation and form potentially carcinogenic aromatic amines via reduction of the azo groups. Furthermore, they can produce potentially hazardous products by photochemical degradation or metabolic processes in plants and animals. The oxidative processes of enzymatic treatment with laccase and ultrasound treatment that leads to dye degradation were monitored by IPC [114]. The IPC separation of aromatic carboxylic acids in coal oxidation products was complemented by an extensive retention behavior study [115,116].

Chapter 7 focused on ionic liquids (ILs) as novel and tunable IPRs. ILs were considered environmentally friendly, but full toxicity studies are in progress [117] and highlight significant ecotoxicities of many ionic liquids [118]. The exponentially increasing use of ILs produced on an industrial scale parallels the increasing number of applications dealing with their analysis and control. To increase RP-HPLC retention of ILs, ion-pairing interactions may be useful. Chaotropic IPRs such as $NaPF_6$ or $NaClO_4$ enhanced their peak shapes and increased their retention [119]. IPC with chaotropic IPR was found to be a highly effective strategy for separating ILs even if the chaotropic IPR did not differentiate cations associated with different anions [120].

Speciation and analysis of many heavy metals and organometallic species of environmental concern are important environmental issues because the toxicity of an element often depends on its oxidation state and chemical architecture; this issue was discussed above (see Section 13.1 and References 2 through 18).

13.6   ENANTIOMERIC SEPARATIONS

Enantioseparation is an important goal for separation scientists. The most common strategy to achieve enantioselectivity is to perform the separation on a chiral column using a chiral selector immobilized onto the chromatographic stationary phase. The two enantiomers are selectively retained based on their different adsorption interactions with the immobilized chiral selector. IPC using chiral IPRs may offer a valid alternative to chiral stationary phases for enantioseparations.

Chiral surfactants, introduced by Terabe in 1989 [121] may yield a pseudo-stationary phase [122–124]. The two enantiomers are separated because of the differences in equilibrium constants for the complex formation with the chiral selector in the mobile and stationary phases and differences in adsorption properties of the formed diastereomeric duplex [125]. The IPR may provide the three-point interaction with the analyte that is necessary for chiral recognition [126]. This field of research is challenging because expensive chiral stationary phases are not widely available. Developing an enantioseparative method still requires a great deal of trial and error. Because the success of a chiral IPC depends on how long-lived the diastereomeric ion-pairs are, the acid–base behavior and hydrophobicity of the chiral IPR assume paramount importance. Karlsson and coworkers [127] found that the best molecular descriptor for enantioseparation is the averaged non-polar unsaturated surface area of the analyte–chiral IPR complex, and the largest separation factors were observed at high IPR concentration.

The most recent enantioseparation relies on a porous graphitic carbon (PGC) stationary phase consisting of well-organized benzene rings that provide a π-π electron-rich very flat surface. This ability to resolve difficult-to-separate isomers sets PGC apart among chromatographic packings. Z-L-alanyl-L-glutaric acid (Z-L-ala-L-glu), Z-L-phenylanyl-L-glutaric acid (Z-L-phe-L-glu), and Z-glycyl-L-glutaric acid (Z-gly-L-glu) peptides were recently tested as chiral counter ions for enantiomeric resolution of amino alcohols. Both the charge and structure of the chiral counter ion and solute influenced enantioselective retention as expected on the basis of a mechanism of complex formation different from the simple Bjerrum type of ion-pair (see Chapter 2) [128].

The simultaneous presence of a chiral selector and a charged non-chiral IPR was studied successfully [129]. The presence of a non-chiral IPR dramatically improved the separation of oppositely charged compounds on a chiral column, probably because the IPR increased retention and hence interactions with the chiral packing, as in the speciation of selenium-containing amino acids, on a crown ether column [5]. IPR adsorption onto a stationary phase is key for enhancing selectivity. When the additives were discontinued, their positive effect did not disappear instantaneously and a memory effect was noted [130].

REFERENCES

1. Cecchi, T. Ion-pairing. *Crit. Rev. Anal. Chem.* 2008, 38, 1–53.
2. Srijaranai, S. et al. Flow-injection in-line complexation for ion-pair reversed phase high performance liquid chromatography of some metal-4-(2-pyridylazo) resorcinol chelates. *Talanta* 2006, 68, 1720–1725.
3. Wolf, R.E., Morrison, J.M., and Goldhaber, M.B. Simultaneous determination of Cr(iii) and Cr(vi) using reversed-phased ion-pairing liquid chromatography with dynamic reaction cell inductively coupled plasma mass spectrometry. *J. Anal. Atom. Spectrom.* 2007, 22, 1051–1060.
4. Helfrich, A. and Bettmer, J. Determination of phytic acid and its degradation products by ion-pair chromatography (IPC) coupled to inductively coupled plasma-sector field-mass spectrometry (ICP-SF-MS). *J. Anal. Atom. Spectrom.* 2004, 19, 1330–1334.

5. Kapolna, E. et al. Selenium speciation studies in Se-enriched chives (*Allium schoenoprasum*) by HPLC-ICP–MS. *Food Chem.* 2007, 101, 1398–1406.

6. Vonderheide, A.P. et al. Investigation of selenium-containing root exudates of *Brassica juncea* using HPLC-ICP-MS and ESI-qTOF-MS. *Analyst* 2006, 131, 33–40.

7. Loreti, V. et al. Biosynthesis of Cd-bound phytochelatins by *Phaeodactylum tricornutum* and their speciation by size-exclusion chromatography and ion-pair chromatography coupled to ICP-MS. *Anal. Bioanal. Chem.* 2005, 383, 398–403.

8. Richardson, D.D., Sadi, B.B.M., and Caruso, J.A. Reversed phase ion-pairing HPLC-ICP-MS for analysis of organophosphorus chemical warfare agent degradation products. *J. Anal. Atom. Spectrom.* 2006, 21, 396–403.

9. Grotti, M. et al. Arsenobetaine is a significant arsenical constituent of the red Antarctic alga *Phyllophora antarctica*. *Environ. Chem.* 2008, 5, 171–175.

10. Yathavakilla, S.K.V. and Caruso, J.A. A study of Se-Hg antagonism in *Glycine max* (soybean) roots by size exclusion and reversed phase HPLC-ICPMS. *Anal. Bioanal. Chem.* 2007, 389, 715–723.

11. Vallant, B., Kadnar, R., and Goessler, W. Development of a new HPLC method for determination of inorganic and methylmercury in biological samples with ICP-MS detection. *J. Anal. Atom. Spectrom.* 2007, 22, 322–325.

12. Wolf, R.E., Morrison, J.M., and Goldhaber, M.B. Simultaneous determination of Cr(iii) and Cr(vi) using reverse-phased ion-pairing liquid chromatography with dynamic reaction cell inductively coupled plasma mass spectrometry. *J. Anal. Atom. Spectrom.* 2007, 22, 1051–1060.

13. Pan, F., Tyson, J.F., and Uden, P.C. Simultaneous speciation of arsenic and selenium in human urine by high-performance liquid chromatography inductively coupled plasma mass spectrometry. *J. Anal. Atom. Spectrom.* 2007, 22, 931–937.

14. Zheng, J. Shibata, Y. Tanaka, A Study of the stability of selenium compounds in human urine and determination by mixed ion-pair reversed-phase chromatography with ICP-MS detection. *Anal. Bional. Chem.* 2002, 374, 348–353.

15. Qiu, J. et al. On-line pre-reduction of Se(VI) by thiourea for selenium speciation by hydride generation. *Spectrochim. Acta B.* 2006, 61, 803–809.

16. Sarica, D.Y., Turker, A.R., and Erol, E. On-line speciation and determination of Cr(III) and Cr(VI) in drinking and waste water samples by reversed-phase high performance liquid chromatography coupled with atomic absorption spectrometry. *J. Sep. Sci.* 2006, 29, 1600–1606.

17. Ko, F.H., Chen, S.L., and Yang, M.H. Evaluation of the gas–liquid separation efficiency of a tubular membrane and determination of arsenic species in groundwater by liquid chromatography coupled with hydride generation atomic absorption spectrometry. *J. Anal. Atom. Spectrom.* 1997, 12, 589–595.

18. Shade, C.W. Automated simultaneous analysis of monomethyl and mercuric Hg in biotic samples by Hg–thiourea complex liquid chromatography following acidic thiourea leaching. *Environ. Sci. Technol.* 2008, 42, 6604–6610.

19. Aguilar, F.J.A. et al. Analytical speciation of chromium in in-vitro cultures of chromate-resistant filamentous fungi. *Anal. Bioanal. Chem.* 2008, 390, 1–8.

20. Shen, X. et al. Investigation of copper–azomacrocyclic complexes by high-performance liquid chromatography. *Biomed. Chromatogr.* 2006, 20, 37–47.

21. Threeprom, J. et al. Simultaneous determination of Cr(III) and Cr(VI) with pre-chelation of Cr(III) using phthalate by ion interaction chromatography with a C18 column. *Talanta* 2007, 71, 103–108.

22. Garcia-Marco, S., Torreblanca, A., and Lucena, J.J. Chromatographic determination of Fe chelated by ethylenediamine-N-(o-hydroxyphenylacetic)-N'-(p-hydroxyphenylacetic) acid in commercial EDDHA/Fe^{3+} fertilizers. *J. Agr. Food Chem.* 2006, 54, 1380–1386.

23. Souza e Silva, R. et al. Separation and determination of metallocyanide complexes of Fe(II), Ni(II) and Co(III) by ion-interaction chromatography with membrane suppressed conductivity detection applied to analysis of oil refinery streams (sour water). *J. Chromatogr. A.* 2006, 1127, 200–206.

24. Jaison, P.G., Raut, N.M., and Aggarwal, S.K. Direct determination of lanthanides in simulated irradiated thoria fuels using reversed-phase high-performance liquid chromatography. *J. Chromatogr. A.* 2006, 1122, 47–53.

25. Soukup-Hein, R.J. et al. A general, positive ion mode ESI-MS approach for the analysis of singly charged inorganic and organic anions using a dicationic reagent. *J. Anal. Chem.* 2007, 79, 7346–7352.

26. Chambers, S.D. and Lucy, C.A. Surfactant coated graphitic carbon based stationary phases for anion-exchange chromatography. *J. Chromatogr. A.* 2007, 1176, 178–184.

27. Harrison, C.R, Sader, J.A., and Lucy C.A. Sulfonium and phosphonium, new ion-pairing agents with unique selectivity toward polarizable anions. *J. Chromatogr. A,* 2006, 1113, 123–129.

28. Kesiuūnaite, G. et al. Ion-pair chromatographic monitoring of photographic effluents. *Centr. Eur. J. Chem.* 2007, 5, 455–465.

29. Fontannaz, P., Kilinc, T., and Heudi, O. HPLC-UV determination of total vitamin C in a wide range of fortified food products. *Food Chem.* 2006, 94, 626–631.

30. Calbiani, F. et al. Validation of an ion-pair liquid chromatography-electrospray-tandem mass spectrometry method for the determination of heterocyclic aromatic amines in meat-based infant foods. *Food Addit. Contam.* 2007, 24, 833–841.

31. Hlabangana, L., Hernandez-Cassou, S., and Saurina J. Determination of biogenic amines in wines by ion-pair liquid chromatography and post-column derivatization with 1,2-naphthoquinone-4-sulphonate. *J. Chromatog. A.* 2006, 1130, 130–136.

32. Lavizzari, T. et al. Improved method for the determination of biogenic amines and polyamines in vegetable products by ion-pair high-performance liquid chromatography. *J. Chromatog. A.* 2006, 1129, 67–72.

33. Castro-Rubio, A. et al. Determination of soybean proteins in soybean–wheat and soybean–rice commercial products by perfusion reversed phase high-performance liquid chromatography. *Food Chem.* 2007, 100, 948–955.

34. Huang, Z. et al. Determination of cyclamate in foods by high performance liquid chromatography–electrospray ionization mass spectrometry. *Anal. Chim. Acta* 2006, 555, 233–237.

35. Wybraniec, S. Effect of tetraalkylammonium salts on retention of betacyanins and decarboxylated betacyanins in ion-pair reversed-phase high-performance liquid chromatography. *J. Chromatog. A.* 2006, 1127, 70–75.

36. Fletouris, D.J. and Papapanagiotou, E.P. A new liquid chromatographic method for routine determination of oxytetracycline marker residue in the edible tissues of farm animals. *Anal. Bioanal. Chem.* 2008, 391, 1189–1198.

37. Cherlet, M., De Baere, S., and De Backer, P. Quantitative determination of dihydrostreptomycin in bovine tissues and milk by liquid chromatography–electrospray ionization–tandem mass spectrometry. *J. Mass Spectrom.* 2007, 42, 647–656.

38. Viñas, P., Balsalobre, N., and Hernández-Córdoba, M. Liquid chromatography on an amide stationary phase with post-column derivatization and fluorimetric detection for determination of streptomycin and dihydrostreptomycin in foods. *Talanta* 2007, 72, 808–812.

39. Amigo-Benavent, M., Villamiel, M., and Castillo, M. Chromatographic and electrophoretic approaches for the analysis of protein quality of soy beverages. *J. Sep. Sci.* 2007, 30, 502–507.

40. Delgado-Andrade, C., Rufián-Henares, J.A., and Morales, F.J. Lysine availability is diminished in commercial fibre-enriched breakfast cereals. *Food Chem.* 2007, 100, 725–731.

41. Oguri, S., Enami, M., and Soga, N. Selective analysis of histamine in food by means of solid-phase extraction cleanup and chromatographic separation. *J. Chromatog. A.* 2007, 1139, 70–74.

42. Cordell, R.L. et al. Quantitative profiling of nucleotides and related phosphate-containing metabolites in cultured mammalian cells by liquid chromatography tandem electrospray mass spectrometry. *J. Chromatog. B.* 2008, 871, 115–124.

43. Yeung, P., Ding, L., and Casley, W.L. HPLC assay with UV detection for determination of RBC purine nucleotide concentrations and application for biomarker study in vivo. *J. Pharm. Biomed.* 2008, 47, 377–382.

44. Timmons, S.C. and Jakeman, D.L. Stereospecific synthesis of sugar-1-phosphates and their conversion to sugar nucleotides. *Carbohydr. Res.* 2008, 343, 865–874.

45. Olesen, K.M. et al. Determination of leukocyte DNA 6-thioguanine nucleotide levels by high-performance liquid chromatography with fluorescence detection. *J. Chromatog. B.* 2008, 864, 149–155.

46. Pruvost, A. et al. Specificity enhancement with LC-positive ESI-MS/MS for the measurement of nucleotides: application to the quantitative determination of carbovir triphosphate, lamivudine triphosphate and tenofovir diphosphate in human peripheral blood mononuclear cells. *J. Mass Spectrom.* 2008, 43, 224–233.

47. Nesterenko, E.P. et al. Separation of nucleic acid precursors on an amphoteric surfactant modified monolith using combined eluent flow, pH and concentration gradient. *J. Sep. Sci.* 2007, 30, 2910–2916.

48. Nesterova, I.V. et al. Metallo-phthalocyanine near-IR fluorophores: oligonucleotide conjugates and their applications in PCR assays. *Bioconjugate Chem.* 2007, 18, 2159–2168.

49. Gill, B.D. and Indyk, H.E. Determination of nucleotides and nucleosides in milks and pediatric formulas: a review. *J. AOAC Int.* 2007, 90, 1354–1364.

50. Xiong, W. et al. Separation and sequencing of isomeric oligonucleotide adducts using monolithic columns by ion-pair reversed-phase nano-HPLC coupled to ion trap mass spectrometry. *Anal. Chem.* 2007, 79, 5312–5321.

51. Chen, Y. et al. Development of an ion-pair HPLC method for investigation of energy charge changes in cerebral ischemia of mice and hypoxia of Neuro-2a cell line. *Biomed. Chromatogr.* 2007, 21, 628–634.

52. Bisjak, C.P. et al. Novel monolithic poly(phenyl acrylate-co-1,4-phenylene diacrylate) capillary columns for biopolymer chromatography. *J. Chromatog. A.* 2007, 1147, 46–52.

53. Jakschitz, T.A.E. et al. Monolithic poly[(trimethylsilyl-4-methylstyrene)-co-bis(4-vinylbenzyl)dimethylsilane] stationary phases for the fast separation of proteins and oligonucleotides. *J. Chromatog. A.* 2007, 1147, 53–58.

54. Liao, Q. et al. Investigation of enzymatic behavior of benzonase/alkaline phosphatase in the digestion of oligonucleotides and DNA by ESI-LC/MS. *Anal. Chem.* 2007, 79, 1907–1917.

55. Shen, H. et al. Development of a magnetic nanoparticle-based artificial cleavage reagent for site-selective cleavage of single-stranded DNA. *Chem. Mater.* 2007, 19, 3090–3092.

56. Dickman, M.J. Post-column nucleic acid intercalation for the fluorescent detection of nucleic acids using ion-pair reverse phase high-performance liquid chromatography. *Anal. Biochem.* 2007, 360, 282–287.

57. Jin, C.H. et al. Effect of mobile phase additives on resolution of some nucleic compounds in high performance liquid chromatography. *Biotechnol. Bioproc. E.* 2007, 12, 525–530.

58. Kok, R.M. et al. Global DNA methylation measured by liquid chromatography-tandem mass spectrometry: analytical technique, reference values and determinants in healthy subjects. *Clin. Chem. Lab. Med.* 2007, 45, 903–911.

59. Huang, W.Y. et al. Rapid prenatal diagnosis of Duchesne muscular dystrophy with gene duplications by ion-pair reversed-phase high-performance liquid chromatography coupled with competitive multiplex polymerase chain reaction strategy. *Prenatal Diag.* 2007, 27, 653–656.

60. Oberacher, H. et al. Some guidelines for the analysis of genomic DNA by PCR-LC-ESI-MS. *J. Am. Soc. Mass Spectrom.* 2006, 17, 124–129.

61. Pfeifer, N. et al. Statistical learning of peptide retention behavior in chromatographic separations: a new kernel-based approach for computational proteomics. *BMC Bioinformatics* 2007, 8, 468; http://www.biomedcentral.com/1471-2105/8/468.

62. Spicer, V. et al. Sequence-specific retention calculator: a family of peptide retention time prediction algorithms in reversed-phase HPLC: applicability to various chromatographic conditions and columns. *Anal. Chem.* 2007, 79, 8762–8768.

63. Ding, W., Hill, J.J., and Kelly, J. Selective enrichment of glycopeptides from glycoprotein digests using ion-pairing normal-phase liquid chromatography. *Anal. Chem.* 2007, 79, 8891–8899.

64. Kim, J. et al. Phosphopeptide elution times in reversed-phase liquid chromatography. *J. Chromatog. A.* 2007, 1172, 9–18.

65. Wang, X. and Carr, P.W. Unexpected observation concerning the effect of anionic additives on the retention behavior of basic drugs and peptides in reversed-phase liquid chromatography. *J. Chromatog. A.* 2007, 1154, 165–173.

66. Fanciulli, G. et al. Liquid chromatography-mass spectrometry assay for quantification of gluten exorphin B5 in cerebrospinal fluid. *J. Chromatogr. B.* 2007, 852, 485–490.

67. Conlon, J.M. Purification of naturally occurring peptides by reversed-phase HPLC. *Nat. Protocols* 2007, 2, 191–197.

68. Armstrong, M., Jonscher, K., and Reisdorph, N.A. Analysis of 25 underivatized amino acids in human plasma using ion-pairing reversed-phase liquid chromatography/time-of-flight mass spectrometry. *Rapid Commun. Mass Spectrom.* 2007, 21, 2717–2726.

69. Liu, D.L., Beegle, L.W., and Kanik, I. Analysis of underivatized amino acids in geological samples using ion-pairing liquid chromatography and electrospray tandem mass spectrometry. *Astrobiology* 2008, 8, 229–241.

70. De Person, M., Chaimbault, P., and Elfakir, C. Analysis of native amino acids by liquid chromatography/electrospray ionization mass spectrometry: comparative study between two sources and interfaces. *J. Mass Spectrom.* 2008, 43, 204–215.

71. Gu, L., Jones, A.D., and Last, R.L. LC-MS/MS assay for protein amino acids and metabolically related compounds for large-scale screening of metabolic phenotypes. *Anal. Chem.* 2007, 79, 8067–8075.

72. Eckstein, J.A. et al. Analysis of glutamine, glutamate, pyroglutamate, and GABA in cerebrospinal fluid using ion-pairing HPLC with positive electrospray LC/MS/MS. *J. Neurosci. Meth.* 2008, 171, 190–196.

73. Dernovics, M. and Lobinski, R. Characterization of the selenocysteine-containing metabolome in selenium-rich yeast II: on the reliability of the quantitative determination of selenocysteine. *J. Anal. Atom. Spectrom.* 2008, 23, 744–751.

74. Kaewsrichan, J. et al. Ion-paired RP-HPLC for determining fermentation end products of *Streptococcus mutans* grown under different conditions. *Science Asia* 2008, 34, 193–198.

75. Lu, W., Bennett, B.D., and Rabinowitz, J.D. Analytical strategies for LC-MS-based targeted metabolomics. *J. Chromatog. B.* 2008, 871, 236–242.

76. Seifar, R.M. et al. Quantitative analysis of metabolites in complex biological samples using ion-pair reversed-phase liquid chromatography-isotope dilution tandem mass spectrometry *J. Chromatog. A.* 2008, 1187, 103–110.

77. Luo, B. et al. Simultaneous determination of multiple intracellular metabolites in glycolysis, pentose phosphate pathway and tricarboxylic acid cycle by liquid chromatography mass spectrometry. *J. Chromatog. A.* 2007, 1147, 153–164.

78. Manyanga, V. et al. Improved liquid chromatographic method with pulsed electrochemical detection for the analysis of gentamicin, *J. Chromatog. A.* 2008, 1189, 347–354.

79. Mascher, D.G., Unger, C.P., and Mascher, H.J. Determination of neomycin and bacitracin in human or rabbit serum by HPLC–MS/MS. *J. Pharm. Biomed. Anal.* 2007, 43, 691–700.

80. Lu, C.Y. and Feng, C.H. Micro-scale analysis of aminoglycoside antibiotics in human plasma by capillary liquid chromatography and nanospray tandem mass spectrometry with column switching. *J. Chromatogr. A.* 2007, 1156, 249–253.

81. Venugopal, K. et al. Development and validation of ion-pairing RP-HPLC method for the estimation of gatifloxacin in bulk and formulations. *J. Chromatog. Sci.* 2007, 45, 220–225.

82. Mohammad, M.A. et al. Stability indicating methods for the determination of norfloxacin in mixture with tinidazole. *Chem. Pharm. Bull.* 2007, 55, 1–6.

83. Kallel, L. et al. Optimization of the separation conditions of tetracyclines on a preselected reversed-phase column with embedded urea group. *J. Sep. Sci.* 2006, 29, 929–935.

84. Yang, M.C. et al. Quantitative determination of fosfomycin calcium and its related substances by ion-pair HPLC method. *Pharm. Care Res.* 2007, 7, 343–346.

85. Kyriakides, D. and Panderi I. Development and validation of a reversed-phase ion-pair high-performance liquid chromatographic method for the determination of risedronate in pharmaceutical preparations. *Anal. Chim. Acta* 2007, 584, 153–159.

86. Xie, Z., Jiang, Y., and Zhang, D. Simple analysis of four bisphosphonates simultaneously by reverse phase liquid chromatography using n-amylamine as volatile ion-pairing agent. *J. Chromatogr. A.* 2006, 1104, 173–178.

87. Ruiz-Angel, M.J., Carda-Broch, S., and Berthod, A. Ionic liquids versus triethylamine as mobile phase additives in the analysis of β-blockers, *J. Chromatog. A.* 2006, 1119, 202–208.

88. Hashem, H. and Jira, T. A. Effect of chaotropic mobile phase additives on retention behaviour of beta-blockers on various reversed-phase high-performance liquid chromatography columns. *J. Chromatogr. A.* 2006, 1133, 69–75.

89. Korir, A.K. et al. Ultraperformance ion-pair liquid chromatography coupled to electrospray time-of-flight mass spectrometry for compositional profiling and quantification of heparin and heparan sulfate. *Anal. Chem.* 2008, 80, 1297–1306.

90. Pucciarelli, F., Passamonti, P., and Cecchi, T. Separation of all isomers of pyridinedicarboxylic acids by ion-pairing chromatography. *J. Liq. Chromatogr. Rel. Technol.* 1997, 20, 2233–2240.

91. Krock, B., Seguel, C.G., and Cembella, A.D. Toxin profile of *Alexandrium catenella* from the Chilean coast as determined by liquid chromatography with fluorescence detection and liquid chromatography coupled with tandem mass spectrometry. *Harm. Algae* 2007, 6, 734–744.

92. Mancini, F. et al. Monolithic stationary phase coupled with coulometric detection: development of an ion-pair HPLC method for the analysis of quinone-bearing compounds. *J. Sep. Sci.* 2007, 30, 2935–2942.

93. Yalcın, G. and Yuktas, F.N. Efficient separation and method development for the quantifying of two basic impurities of nicergoline by reversed-phase high performance liquid chromatography using ion-pairing counter ions. *J. Pharm. Biomed. Anal.* 2006, 42, 434–440.

94. Ariffin, M.M. and Anderson, R.A. LC/MS/MS analysis of quaternary ammonium drugs and herbicides in whole blood. *J. Chromatogr. B.* 2006, 842, 91–97.

95. Lee, M. et al. Selective solid-phase extraction of catecholamines by the chemically modified polymeric adsorbents with crown ether, *J. Chromatogr. A.* 2007, 1160, 340–344.

96. Manickum, T. Interferences by anti-TB drugs in a validated HPLC assay for urinary catecholamines and their successful removal. *J. Chromatogr. B.* 2008, 873, 124–128.

97. Putzbach, K. et al. Determination of bitter orange alkaloids in dietary supplements standard reference materials by liquid chromatography with ultraviolet absorbance and fluorescence detection. *J. Chromatogr. A.* 2007, 1156, 304–311.

98. Ding, L. et al. Simultaneous determination of flavonoid and alkaloid compounds in citrus herbs by high-performance liquid chromatography-photodiode array detection-electrospray mass spectrometry. *J. Chromatogr. B.* 2007, 857, 202–209.

99. Bharathi, V.D. et al. LC-MS-MS assay for simultaneous quantification of fexofenadine and pseudoephedrine in human plasma. *Chromatographia* 2008, 67, 461–466.

100. Flieger J. Effect of chaotropic mobile phase additives on the separation of selected alkaloids in reversed-phase high-performance liquid chromatography. *J. Chromatogr. A.* 2006, 1113, 37–44.

101. Fuh, M.R., Wu, T.Y., and Lin, T.Y. Determination of amphetamine and methamphetamine in urine by solid phase extraction and ion-pair liquid chromatography–electrospray–tandem mass spectrometry. *Talanta* 2006, 68, 987–991.

102. Nakajima, R. et al. Direct determination of p-hydroxymethamphetamine glucuronide in human urine by high-performance liquid chromatography. *Chem. Pharm. Bull.* 2006, 54, 493–495.

103. Gao, S. et al. Evaluation of volatile ion-pair reagents for the liquid chromatography–mass spectrometry analysis of polar compounds and its application to the determination of methadone in human plasma. *J. Pharm. Biomed. Anal.* 2006, 40, 679–688.

104. Aramendia, M.A. et al. Determination of diquat and paraquat in olive oil by ion-pair liquid chromatography–electrospray ionization mass spectrometry. *Food Chem.* 2006, 97, 181–188.

105. Moret, S., Hidalgo, M., and Sánchez, J.M. Development of ion-pairing liquid chromatography method for the determination of phenoxyacetic herbicides and their main metabolites: application to the analysis of soil samples. *Chromatographia* 2006, 63, 109–115.

106. Marín, J.M. et al. An ion-pairing liquid chromatography/tandem mass spectrometric method for the determination of ethephon residues in vegetables. *Rapid Commun. Mass Spectrom.* 2006, 20, 419–426.

107. Quintana, J.B., Rodil, R., and Reemtsma, T. Determination of phosphoric acid mono- and diesters in municipal wastewater by solid-phase extraction and ion-pair liquid chromatography–tandem mass spectrometry. *Anal. Chem.* 2006, 78, 1644–1650.

108. Lin, C.Y. and Huang, S.D. Application of liquid–liquid–liquid microextraction and ion-pair liquid chromatography coupled with photodiode array detection for the determination of chlorophenols in water. *J. Chromatogr. A.* 2008, 1193, 79–84.

109. Esser, S., Gabelmann, H., and Wenclawiak, B.W. High performance liquid chromatography for the determination of chelating agents in waste water. *Anal. Lett.* 2007, 40, 1811–1819.

110. Quintana, J.B. and Reemtsma, T. Rapid and sensitive determination of ethylenediamine tetraacetic acid and diethylenetriaminepentaacetic acid in water samples by ion-pair reversed-phase liquid chromatography–electrospray tandem mass spectrometry. *J. Chromatogr. A.* 2007, 1145, 110–117.

111. Richardson, D.D., Sadi, B.B.M., and Caruso, J.A. Reversed phase ion-pairing HPLC-ICP-MS for analysis of organophosphorus chemical warfare agent degradation products. *J. Anal. Atom. Spectrom.* 2006, 21, 396–403.

112. Jin, F. et al. Determination of diallyldimethylammonium chloride in drinking water by reversed-phase ion-pair chromatography-electrospray ionization mass spectrometry. *J. Chromatogr. A.* 2006, 1101, 222–225.

113. Chen, H.C., Wang, S.P., and Ding, W.H. Determination of fluorescent whitening agents in environmental waters by solid-phase extraction and ion-pair liquid chromatography–tandem mass spectrometry. *J. Chromatogr. A.* 2006, 1102, 135–142.

114. Tauber, M.M., Gubitz, G.M., and Rehorek, A.Degradation of azo dyes by oxidative processes: laccase and ultrasound treatment. *Biores. Technol.* 2008, 99, 4213–4220.

115. Kawamura, K. et al. Separation of aromatic carboxylic acids using quaternary ammonium salts on reversed-phase HPLC 1: separation behavior of aromatic carboxylic acids. *Sep. Sci. Technol.* 2006, 41, 379–390.

116. Kawamura, K. et al. Separation of aromatic carboxylic acids using quaternary ammonium salts on reversed-phase HPLC 2: application for the analysis of Loy Yang coal oxidation products. *Sep. Sci. Technol.* 2006, 41, 723–732.

117. Studzinńska, S., Stepnowski, P., and Buszewski, B. Chromatographic and chemometric methods for evaluation of properties of ionic liquids. *Acta Chim. Slov.* 2007, 54, 20–24.

118. Zhao, D., Liao, Y., and Zhang, Z. Toxicity of Ionic Liquids. *Clean,* 2007, 35, 42–48.

119. Ruiz-Angel, M.J. and Berthod, A. Poster P01-10 presented at 31st International Symposium on High Performance Liquid Phase Separations and Related Techniques, Ghent, June 2007.

120. Ruiz-Angel, M.J. and Berthod, A.Reversed-phase liquid chromatography analysis of alkyl-imidazolium ionic liquids II: effects of different added salts and stationary phase influence. *J. Chromatogr. A.* 2008, 1189, 476–482.

121. Terabe, S., Shibata, M., and Miyashita Y. Chiral separation by electronkinetic chromatography while bile salt micelles. *J. Chromatogr. A.* 1989, 480, 403–411.

122. Thibodeaux, S.J., Billiot, E., and Warner, I.M. Enantiomeric separations using poly(l-valine) and poly(l-leucine) surfactants: investigation of steric factors near chiral center. *J. Chromatog. A.* 2002, 966, 179–186.

123. Billiot, F.H., Billiot, E.J., and Warner, I.M. Depth of penetration of binaphthyl derivatives into the micellar core of sodium undecenoyl leucyl-leucinate surfactants. 2002, 950, 233–239.

124. Nimura, N. et al. Chiral mobile phase additives. In *Chiral Separations by HPLC*, Krstulovic, A.M., Ed. Ellis Horwood: Chichester, 1989, pp. 107–172.

125. Karlsson, A., and Pettersson, C. Enantiomeric separation of amines using N-benzoxycarbonylglycyl-proline as chiral additive and porous graphitic carbon as solid phase. *J. Chromatogr. A.* 1991, 543, 287–297.

126. Lochmuller, C.H., and Souter, R.W. Chromatographic resolution of enantiomers: Selective review. *J. Chromatogr.* 1975, 113(3), 283–302.

127. Karlsson, A. et al. Enantioselective Ion-Pair Chromatography of Phenolic 2-Dipropylaminotetralin Derivatives on Achiral Stationary Phases: an Experimental and Theoretical Study of Chiral Discrimination. *Acta Chem. Scand.* 1993, 47, 469–481.

128. Karlsson, A. and Almgren, K. Reversal of enantiomeric retention order by using a single N-derivatized dipeptide as chiral mobile phase additive and porous graphitised carbon as stationary phase. *Chromatographia* 2007, 66, 349–356.

129. Sun, S. et al. Chiral ligand exchange chromatography for separation of three stereoisomers of octahydroindole-2-carboxylic acid. *Chromatographia* 2006, 63, 331–335.

130. Ye, Y.K., Lord, B., and Stringham, R.W. Memory effect of mobile phase additives in chiral separations on a Chiralpak AD column. *J. Chromatogr. A.* 2002, 945, 139–146.

131. Liu, Y.Q. et al. Quantitative determination of erythromycylamine in human plasma by liquid chromatography mass spectrometry and its application in a bioequivalence study of dirithromycin. *J. Chromatog. B.* 2008, 864, 1–8.

132. Mino, Y. et al. Simultaneous determination of mycophenolic acid and its glucuronides in human plasma using isocratic ion-pair high-performance liquid chromatography. *J. Pharm. Biomed.* 2008, 46, 603–608.

133. Chen, X. et al. Simultaneous determination of amodiaquine and its active metabolite in human blood by ion-pair liquid chromatography-tandem mass spectrometry. *J. Chromatog. B.* 2007 860, 18–25.

134. Lefebvre, I. et al. Quantification of zidovudine and its monophosphate in cell extracts by on-line solid-phase extraction coupled to liquid chromatography. *J. Chromatog. B.* 2007, 858, 2–7.

135. Sora, D.I. et al. Validated ion-pair liquid chromatography/fluorescence detection method for assessing the variability of the loratadine metabolism occurring in bioequivalence studies. *Biomed. Chromatogr.* 2007, 21, 1023–1029.

136. Keski-Rahkonen, P. et al. Quantitative determination of acetylcholine in microdialysis samples using liquid chromatography/atmospheric pressure spray ionization mass spectrometry. *Rapid Commun. Mass Spectrom.* 2007, 21, 2933–2943.

137. Tak, V. et al. Application of Doehlert design in optimizing the determination of degraded products of nerve agents by ion-pair liquid chromatography electrospray ionization tandem mass spectrometry. *J. Chromatog. A.* 2007, 1161, 198–206.

138. Oostendorp, R.L. et al. Determination of imatinib mesylate and its main metabolite (CGP74588) in human plasma and murine specimens by ion-pairing reversed-phase high-performance liquid chromatography. *Biomed. Chromatogr.* 2007, 21, 747–754.

139. Lan, K. et al. Quantitative determination of rosuvastatin in human plasma by ion-pair liquid-liquid extraction using liquid chromatography with electrospray ionization tandem mass spectrometry. *J. Pharm. Biomed.* 2007, 44, 540–546.

140. Gao, L. et al. Simultaneous quantification of malonyl-CoA and several other short-chain acyl-CoAs in animal tissues by ion-pairing reversed-phase HPLC/MS. *J. Chromatog. B.* 2007, 853, 303–313.

141. Wang, G. et al. Ultra-performance liquid chromatography/tandem mass spectrometric determination of diastereomers of SCH 503034 in monkey plasma. *J. Chromatog. B.* 2007, 852, 92–100.

142. Kuśmierek, K. and Bald, E. Simultaneous determination of tiopronin and d-penicillamine in human urine by liquid chromatography with ultraviolet detection. *Anal. Chim. Acta* 2007, 590, 132–137.

143. Madan, J. et al. Ion-pairing RP-HPLC analytical methods for simultaneous estimation of simvastatin and its α-hydroxy acid. *J. Sci. Ind. Res.* 2007, 66, 371–376.

144. Guo, Z.Y. et al. HPLC method for the determination of ethacridine lactate in human urine. *Biomed. Chromatogr.* 2007, 21, 480–483.

145. Ji, D. et al. Determination of chondroitin sulfate content in raw materials and dietary supplements by high-performance liquid chromatography with ultraviolet detection after enzymatic hydrolysis: single-laboratory validation. *J. AOAC Int.* 2007, 90, 659–669.

146. Vela, J.E. et al. Simultaneous quantitation of the nucleotide analog adefovir, its phosphorylated anabolites and 2'-deoxyadenosine triphosphate by ion-pairing LC/MS/MS. *J. Chromatog. B.* 2007, 848, 335–343.

147. Hsieh, Y., Duncan, C.J.G., and Brisson, J.M. Porous graphitic carbon chromatography/tandem mass spectrometric determination of cytarabine in mouse plasma. *Rapid Commun. Mass Spectrom.* 2007, 21, 629–634.

148. Kotzagiorgis, E.C., Michaleas, S., and Antoniadou-Vyza, E. Improved photostability indicating ion-pair chromatography method for pergolide analysis in tablets and in the presence of cyclodextrins. *J. Pharm. Biomed.* 2007, 43, 1370–1375.

14 IPC versus Competitive Techniques

IPC is a valuable addition to the toolbox of analytical chemistry for dealing with multi-faceted separations of ionizable compounds. The aim of this chapter is to perform a comparative evaluation of IPC and possible competitive analytical strategies (see Figure 13.1). No significant statistical differences were usually found among results obtained from IPC in comparison to other techniques.

IPC is often compared with capillary electrophoresis (CE) because they share the ionic natures of their putative analytes. For the analysis of furosine, a highly valuable indicator of food quality, IPC and capillary zone electrophoresis (CZE) techniques were employed to determine furosine content in soy-based beverages. Results obtained by both analytical techniques did not differ significantly, confirming their feasibility for furosine analysis [1]. The abilities of IPC and CZE were compared for the analysis of organic acids. Extreme stability, reproducibility, and linearity were the positive attributes of IPC, while CE showed better recovery [2]. Conversely, when both these techniques were compared for the determination of EDTA and other complexing agents in cosmetic products, the concentrations found were comparable, but IPC displayed lower detection limit. The advantages of CE were shorter analysis time and less chemical consumption [3].

IPC usually offers better efficiency than IC. In an analytical speciation of chromium in in-vitro cultures of chromate-resistant filamentous fungi, IPC and anion exchange chromatography gave consistent results [4]. The separation of cytarabine from endogenous compounds in mouse plasma by packed-column supercritical fluid chromatography on a bare silica stationary phase with an isocratic mobile phase of CO_2 and methanol solvent with the addition of ammonium acetate as ionic modifier yielded results consistent with those of IPC [5]. Analytical data from IPC and conventional GC proved consistent in the separation of aromatic carboxylic acids [6] and for determining complexing agents [7]. Obviously, the IPC strategy avoided the derivatization step required to obtain sufficient volatility of analytes.

In a quality evaluation strategy for multi-source active pharmaceutical ingredient starting materials, IPC was viewed as a stronger candidate among several analytical methods because of ease of operation [8]. When a coupled flow injection–IPC method was used to analyze metal complexes in chrome plating wastewater, results agreed well with those of atomic absorption spectrometry (AAS) [9]. Analogous results were produced when IPC was compared to enzyme multiplied immunoassay technique for the determination of mycophenolic acid in plasma [10].

IPC and HILIC were systematically compared to study their metabolomics potential, using both compound standards and cellular extracts under identical mass spectrometry conditions. IPC generally offered better separation and signalling for

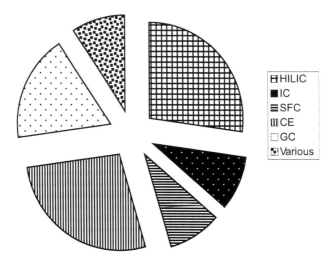

FIGURE 14.1 Comparison of relative importance of IPC and competitive techniques from 2004 to 2008.

negatively charged metabolites while it did not work well in positive ionization mode due to poor retention of amine-containing compounds and ion suppression effects by cationic volatile amine IPRs [11]. HILIC served as a useful alternative to IPC for simultaneous separation of six impurities of mildronate [12]. In the comparison of methods for the analysis of central carbon metabolites, IPC-MS was particularly suited to polar analytes even if IPRs may cause ion suppression and fouling of the instrument. HILIC has the advantage of using solvents that are highly suitable for MS coupling, but retention times are often less reproducible [13].

A cross-validation study showed comparable levels of adefovir and its metabolites determined by IPC-MS/MS or radioactivity detection. The optimized IPC strategy was sufficiently sensitive, accurate, and precise to serve as a useful tool for studying the intracellular pharmacology of adefovir [14].

REFERENCES

1. Amigo-Benavent, M., Villamiel, M., and Dolores del Castillo, M. Chromatographic and electrophoretic approaches for the analysis of protein quality of soy beverages. *J. Sep. Sci.* 2007, 30, 502–507.
2. Wang, S.P. and Liao, C.S. Comparison of ion-pair chromatography and capillary zone electrophoresis for the assay of organic acids as markers of abnormal metabolism. *J. Chromatogr. A.* 2004, 1051, 213–219.
3. Katata, L., Nagaraju, V., and Crouch, A.M. Determination of ethylenediaminetetraacetic acid, ethylenediaminedisuccinic acid and iminodisuccinic acid in cosmetic products by capillary electrophoresis and high performance liquid chromatography. *Anal. Chim. Acta* 2006, 579, 177–184.
4. Aguilar, F.J.A. et al. Analytical speciation of chromium in in-vitro cultures of chromate-resistant filamentous fungi. *Anal. Bioanal. Chem.* 2008, 392, 269–276.

5. Hsieh, Y., Li, F., and Duncan, C.J.G. Supercritical fluid chromatography and high-performance liquid chromatography/tandem mass spectrometric methods for the determination of cytarabine in mouse plasma. *Anal. Chem.* 2007, 79, 3856–3861.

6. Kawamura, K. et al. Separation of aromatic carboxylic acids using quaternary ammonium salts on reversed-phase HPLC 2: application for the analysis of Loy Yang coal oxidation products. *J. Sep. Sci. Technol.* 2006, 41, 723–732.

7. Quintana, J.B., and Reemtsma, T. *J. Chromatogr. A.* Rapid and sensitive determination of ethylenediaminetetraacetic acid and diethylenetriaminepentaacetic acid in water samples by ion-pair reversed-phase liquid chromatography-electrospray tandem mass spectrometry. *J. Chromatogr. A.* 2007, 1145, 110–117.

8. Gavin, P.F. et al. Quality evaluation strategy for multi-sourced active pharmaceutical ingredient (API) starting materials. *J. Pharm. Biomed.* 2006, 41, 1251–1259.

9. Srijaranai, S. et al. Flow-injection in-line complexation for ion-pair reverse phase high performance liquid chromatography of some metal-4-(2-pyridylazo) resorcinol chelates. *Talanta* 2006, 68, 1720–1725.

10. Hosotsubo, H. et al. Analytic validation of the enzyme multiplied immunoassay technique for the determination of mycophenolic acid in plasma from renal transplant recipients compared with a high performance liquid chromatographic assay. *Ther. Drug Monit.* 2001, 23, 669–674.

11. Lu, W., Bennett, B.D., and Rabinowitz, J.D. Analytical strategies for LC-MS-based targeted metabolomics. *J. Chromatogr. B.* 2008, 871, 236–242.

12. Hmelnickis, J. et al. Application of hydrophilic interaction chromatography for simultaneous separation of six impurities of mildronate substance. *J. Pharm. Biomed.* 2008, 48, 649–656.

13. Timischl, B., Dettmer, K., and Oefner, P.J. Mass spectrometry in the analysis of the central carbon metabolism. *Anal. Bioanal. Chem.* 2008, 391, 895–898.

14. Vela, J.E. et al. Simultaneous quantitation of the nucleotide analog adefovir, its phosphorylated anabolites and 2′-deoxyadenosine triphosphate by ion-pairing LC/MS/MS. *J. Chromatogr. B.* 2007 848, 335–343.

15 Ion-Pairing in Different Analytical Techniques

Ion-pairing, as discussed in Chapter 2, exerts dramatic effects on the chemical behaviour of analytes. Changes of charge status and hydrophobicity are the bases for the success of IPC separations as explained in Chapter 3. However, ion-pairing influences many other separative and non-separative techniques. In many papers dealing with theoretical facets of separative techniques, ion-pairing was claimed to profoundly affect separation.

15.1 CAPILLARY ELECTROPHORESIS (CE) AND RELATED TECHNIQUES

CE is an extremely efficient technique and many separation scientists use it because it is practical. CE proved excellent for obtaining ion-pairing equilibrium constants and a thorough description of the ion-pairing process under CE conditions was recently proposed (see Section 2.7.6). Because electrophoretic mobility is directly related to analyte charge status, ionic modifiers in a background electrolyte (BGE) strongly influence CE separation since a decrease of mobility is expected upon ion -pairing. This allows ion-pairing to prevent one of the major drawbacks of CE: the lack of versatility regarding manipulation of separation selectivity for both organic [1–3] and inorganic [4] analytes. Ion-pairing was also used to achieve online sample pre-concentration. As the IPR migrates through the sample zone, it effectively collects oppositely charged analytes into a tightly swept zone [5].

Recent synthesis of chiral ILs paved the way for evaluation of new potential selectors for chiral separations that underscore ion-pairing in CE. Particular selectivities may be achieved by exploiting unique hydrophobic interactions, ion–dipole or ion-induced-dipole, ion-pairing effects and tailoring the molecular architecture of the IL. Even if chiral ILs did not present direct enantioselectivity with regard to model analytes (2-arylpropionic acids), in the presence of classical chiral selectors (di- or trimethyl-β-cyclodextrin), an increase in separation selectivity and resolution suggesting synergistic effects was observed in some cases [6]. Chiral counter ions were exploited in many enantiomeric separations via ion-pairing CE [7,8]

Usually bi- or multi-dimension separations must be run individually since they utilize different physico-chemical separation principles and differ greatly in selectivity. Ion-pairing and electrophoretic mobility provide two compatible separation models that may be run simultaneously in the same single separation device to achieve a bi-dimensional technique. Since they are respectively based on the complex hydrophobicity and on the analyte change, experiments demonstrated that ion-pairing occurs in the BGE and can be optimized in terms of counter ion hydrophobicity and concentration. The major practical repercussion of ion-pairing CE is the high peak capacity that can be obtained [9–11]. This strategy represents a break from the chromatographic resemblance of CE in which the hydrophobic interaction mechanism is at an interface as in capillary electrochromatography (CEC) and capillary micellar electrokinetic chromatography (CMEKC). Ion-pairing also proved valuable in CMEKC which makes use of micellar mobile phases in reversed phase mode. Complex electrostatic hydrophobic and steric interactions exist between the solute and both stationary and mobile phases. The presence of chiral surfactants in the mobile phase afforded enantioseparation also via CMEKC [12]. Ion-pairing focusing of polyanionic heparins in a polycationic polyacrylamide gel, made by incorporating a gradient of positively charged monomers into the neutral polycrylamide backbone allows the polydisperse heparins to reach a steady-state position along the migration path and condense in an environment inducing charge neutralization. Both size and charge distribution along the oligosaccharide chains influence the separations there by confirming the multiplicity of phenomena involved in ion-pairing [13].

15.2 SUPERCRITICAL FLUID CHROMATOGRAPHY (SFC) AND OTHER TECHNIQUES

Several scientific reports about SFC indicate that the chromatographic retention mechanisms of charged analytes in the presence of suitable ionic modifiers involve ion-pairing [14]. Ion-pairing of sulfonates with ammonium salt additives was effectively exploited to enhance the solvating power of the mobile phase [15] and sharpen analyte peaks [16]. The use of ammonium acetate produced unique results (see Figure 15.1). Ion-pairing also explained enantioselectivity when chiral analytes were analyzed with packed columns in the presence of chiral counter ions [17]. An achiral IPR under SFC conditions played a crucial role in the enantioseparation of a variety of amines [18] for reason explained in Section 13.6.

The first preparative fractionation of betalaine pigments by means of ion-pair high-speed countercurrent chromatography utilized a solvent system composed of trifluoroacetic acid at low concentration as the IPR that considerably improved affinity of polar betacyanins and betaxanthins to the organic stationary phase of the biphasic solvent mixture [19].

15.3 UV-VISIBLE SPECTROPHOTOMETRY

Colored ion-pair complexes were found useful for extractive spectrophotometric assays of a number of analytes: oseltamivir [20], nortriptyline hydrochloride in pharmaceutical formulations [21], zolmitriptan in tablets [22], finasteride in tablets [23], and dosage forms of amoxycillin and flucloxacillin [24]. In all cases, the chromogenic reagent and analyte formed ion-pairs that obeyed Beer's law and were suitable for quantitative determinations.

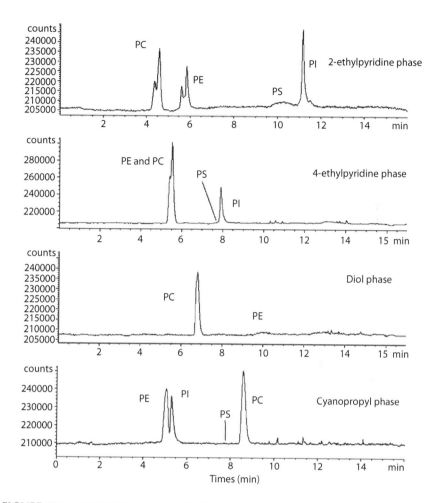

FIGURE 15.1 SFC-ELSD of phospholipid mixture with modified CO_2 on the four stationary phases. The modifier consisted of methanol with 5 mM ammonium acetate. Chromatographic conditions: flow rate of 2 mL/min, methanol/additive concentration raised from 15 to 55% at 4 min and held for 5 min at 55%. (From Yip, H.S.H. et al. *Chromatographia* 2007, 65, 655–665. With permission of Vieweg Verlag.)

15.4 EXTRACTION AND SAMPLE PREPARATION

The transfer of ions as pairs from one phase to a second is important in pre-analytical extractive strategies. Ion-pair extraction of amines with bis-2-ethylhexylphosphate and derivatization with isobutyl chloroformate prior to GC-MS analysis was expedient for the simultaneous determination of aliphatic and aromatic amines in ambient air and airborne particulate matters [25]. The determination of 5-aminoimidazole-4-carboxamide in human plasma relied on the formation and extraction of ion-pairs between the cationic analyte and 1-pentanesulfonate followed by LC-MS/MS [26].

In an ion-pair solid phase extraction (IP-SPE) method for the analysis of alkylphosphonic acids present in aqueous samples, the sample was mixed with a phenyltrimethylammonium hydroxide IPR and passed through an activated charcoal SPE cartridge. The retained complexes in the cartridge were eluted with methanol and analyzed by GC-MS [27]. After treatment with acetic acid and tetrabutylammonium hydroxide, rosuvastatin was extracted from human plasma by simple ion-pair one-step liquid–liquid extraction with an internal standard [28]. Linear alkylbenzene sulfonates and sulfophenylcarboxylic acids were extracted from aqueous environmental samples using methylene green (MG) as the ion-pairing reagent, and they were subsequently derivatized and analyzed by GC-MS [28]. When dealing with ion-pair extraction, the magnitude of the distribution ratio determines the extractability of the analyte; selectivity toward putative interferents should also be considered.

REFERENCES

1. Van Eeckhaut, A. and Michotte, Y. Chiral separations by capillary electrophoresis: recent developments and applications. *Electrophoresis* 2006, 27, 2880–2895.
2. Ma, H. et al. Hydrogen-bond effect and ion-pair association in the separation of neutral calix[4]pyrroles by nonaqueous capillary electrophoresis. *J. Chromatogr. A.* 2008, 1188, 57–60.
3. Steiner, S.A., Watson, D.M., and Fritz J.S. Ion association with alkylammonium cations for separation of anions by capillary electrophoresis. *J. Chromatogr. A.* 2005, 1085, 170–175.
4. Gong, M. et al. Online sample preconcentration by sweeping with dodecyltrimethylammonium bromide in capillary zone electrophoresis, *J. Chromatogr. A.* 2006, 1125, 263–269.
5. François, Y. et al. Evaluation of chiral ionic liquids as additives to cyclodextrins for enantiomeric separations by capillary electrophoresis. *J. Chromatogr. A.* 2007, 1155, 134–141.
6. Kodama, S. et al. Direct chiral resolution of tartaric acid by ion-pair capillary electrophoresis using an aqueous background electrolyte with (1R,2R)-(-)-1,2-diaminocyclohexane as a chiral counterion. *Electrophoresis* 2003, 24, 2711–2715.
7. Loden, H. et al. Development of a chiral non-aqueous capillary electrophoretic system using the partial filling technique with UV and mass spectrometric detection. *J. Chromatogr. A.* 2003, 986, 143–152.
8. Barták, P. et al. Advanced statistical evaluation of the complex formation constants from electrophoretic data II: diastereomeric ion-pairs of (R,S)-N-(3,5- dinitrobenzoyl)leucine and tert-butylcarbamoylquinine. *Anal. Chim. Acta* 2004, 506, 105–113.
9. Popa, T.V., Mant, C.T., and Hodges, R.S. Ion interaction–capillary zone electrophoresis of cationic proteomic peptide standards. *J. Chromatogr. A.* 2006, 1111, 192–199.
10. Popa, T.V. et al. Capillary zone electrophoresis of α-helical diastereomeric peptide pairs with anionic ion-pairing reagents. *J. Chromatogr. A.* 2004, 1043, 113–122.
11. Popa, T.V., Mant, C.T., and Hodges, R.S. Capillary electrophoresis of cationic random coil peptide standards: effect of anionic ion-pairing reagents and comparison with reversed-phase chromatography. *Electrophoresis* 2004, 25, 1219–1229.
12. Wang, H. et al.Sodium maleopimaric acid as pseudostationary phase for chiral separations of amino acid derivatives by capillary micellar electrokinetic chromatography. *J. Sep. Sci.* 2007, 30, 2748–2753.
13. Zilberstein, G. et al. Focusing of Low-Molecular-Mass Heparins in Polycationic Polyacrylamide Matrices. *Anal. Chem.* DOI: 10.1021/ac901050q, In press.

14. Yip, H.S.H. Ashraf-Khorassani, M., and Taylor, L.T. Feasibility of phospholipid separation by packed column SFC with mass spectrometric and light scattering detection. *Chromatographia* 2007, 65, 655–665.

15. Zheng, J. et al. Study of the elution mechanism of sodium aryl sulfonates on bare silica and a cyano-bonded phase with methanol-modified carbon dioxide containing an ionic additive *J. Chromatogr. A.* 1090, 2005, 155–164.

16. Zheng, J., Taylor, L.T., and Pinkston, J.D. Elution of cationic species with/without ion-pair reagents from polar stationary phases via SFC. *Chromatographia* 2006, 63, 267–276.

17. Gyllenhaal, O. and Karlsson, A. Enantioresolution of dihydropyridine substituted acid by supercritical fluid chromatography on Hypercarb® with Z-(L)-arginine as chiral counter ion. *Chromatographia* 2000, 52, 351–355.

18. Stringham, R.W. Chiral separation of amines in subcritical fluid chromatography using polysaccharide stationary phases and acidic additives. *J. Chromatogr. A.* 2005, 1070, 163–170.

19. Jerz, G. et al. Separation of betalains from berries of Phytolacca americana by ion-pair high-speed counter-current chromatography. *J. Chromatogr. A.* 2008, 1190, 63–73.

20. Green, M.D., Nettey, H., and Wirtz, R.A. Determination of oseltamivir quality by colorimetric and liquid chromatographic methods. *Emerg. Infect. Dis.* 2008, 14, 552–556.

21. Misiuk, W. and Tykocka, A. Sensitive extractive spectrophotometric methods for determination of nortriptyline hydrochloride in pharmaceutical formulations. *Chem. Pharm. Bull.* 2007, 55, 1655–1661.

22. Aydogmus, Z. and Inanli, I. Extractive spectrophotometric methods for determination of zolmitriptan in tablets. *J. AOAC Int.* 2007, 90, 1237–1241.

23. Ulu, S.T. A new spectrophotometric method for the determination of finasteride in tablets. *Spectrochim. Acta A.* 2007, 67, 778–783.

24. Aly, H.M. and Amin, A.S. Utilization of ion exchanger and spectrophotometry for assaying amoxycillin and flucloxacillin in dosage form. *Int. J. Pharm.* 2007, 338, 225–230.

25. Akyüz, M. Simultaneous determination of aliphatic and aromatic amines in ambient air and airborne particulate matters by gas chromatography-mass spectrometry. *Atmos. Environ.* 2008, 42, 3809–3819.

26. Chen, X. et al. Determination of 5-aminoimidazole-4-carboxamide in human plasma by ion-pair extraction and LC-MS/MS. *J. Chromatogr. B.* 2007, 855, 140–144.

27. Vijaya-Saradhi, U.V.R. et al. Gas chromatographic-mass spectrometric determination of alkylphosphonic acids from aqueous samples by ion-pair solid-phase extraction on activated charcoal and methylation. *J. Chromatogr. A.* 2007, 1157, 391–398.

28. Lan, K. et al. Quantitative determination of rosuvastatin in human plasma by ion-pair liquid–liquid extraction using liquid chromatography with electrospray ionization tandem mass spectrometry. *J. Pharm. Biomed.* 2007, 44, 540–546.

29. Akyüz, M. Ion-pair extraction and GC-MS determination of linear alkylbenzene sulphonates in aqueous environmental samples. *Talanta* 2007, 71, 471–478.

16 Non-Separative Applications of IPC

This chapter will discuss non-separative applications of IPC in several fields. IPC is a valuable biotechnology for characterizing genomic markers. Denaturing high performance liquid chromatography (DHPLC) is a relatively new IPC technique that allows a chromatographer to determine whether a DNA fragment is identical to a standard fragment or, on the converse, if the DNA contains at least one different nucleotide.

It follows that the technique is useful to identify: (1) point mutation (a single different nucleotide) that may be a silent mutation (also known as a synonymous mutation due to degenerate coding), resulting in a codon that codes for the same amino acid or a missense mutation (type of non-synonymous mutation) producing a codon that codes for a different amino acid; the resulting protein may be non-functional but certain missense mutations can be "quiet" since the protein may still function; polymorphisms, (3) insertions, and deletions.

DHPLC is usually performed on a styrene–divinylbenzene-based polymeric stationary phase, with a mobile phase that contains triethylammonium acetate as the IPR to provide adequate reversed phase (RP) retention for the negatively charged nucleic acid molecules. The samples are usually amplified according to polymerase chain reaction (PCR) protocols and then injected into the chromatographic system.

Figure 16.1 clarifies the way DHPLC works. If the DNA does not contain a mutation after the denaturation (heat) and renaturation (cool) steps, only homoduplexes are formed. Conversely, a heterozygous mutation leads to heteroduplex formation by wild-type and mutated DNA strands during re-annealing (DNA strands in heteroduplexes are not completely correspondent due to a nucleotide mismatch that produces a bubble). Heteroduplexes cannot be formed if a homozygous mutation is present unless non-mutant DNA is added to the sample. Heteroduplex species are separated from homoduplex molecules on a styrene–divinylbenzene polymeric stationary phase with variable heat denaturation of the DNA strands. DNA fragments are usually eluted according to their dimensions during an acetonitrile gradient. In the absence of mutations, only one chromatographic peak (homoduplex) will be detected. If heteroduplexes are formed, four chromatographic peaks will be obtained.

This method provided efficient genotyping [1,2] by detecting single base changes and short deletions or insertions [3,4], polymorphisms [5,6], and point mutations in drug-resistant genes of *Mycobacterium tuberculosis* [7]. A conjugated PCR/IPC technique proved promising for characterizing exonic deletions that are left unexamined in most routine mutation analysis [8].

In DHPLC, ESI-MS detection offers intrinsic advantage over UV detection since the molecular mass of an intact nucleic acid reflects the nucleotide composition (but not the

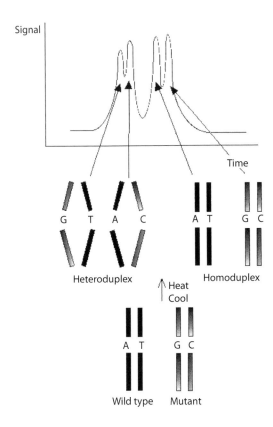

FIGURE 16.1   Steps of denaturing high performance liquid chromatography (DHPLC).

order). The potential of ESI-MS was demonstrated in the simultaneous characterizations of lengths and nucleotide polymorphisms within human mitochondrial DNA. The findings were confirmed by direct sequencing of the PCR products [9]. An hyphenated IPC-ESI-MS technique provided a promising alternative to capillary electrophoresis (CE) of highly polymorphic short tandem repeats for the analysis of allelic imbalance that is of great importance for understanding tumorigenesis and for the clinical management of malignant disease [10]. In conclusion, DHPLC is a valuable tool for comparative nucleic acid sequencing and evolutionary, forensic, and genetic studies.

IPC can also be effective for determining thermodynamic and physical data of a studied system. For example, the study of IPR adsorption isotherms may provide the surface area and the monolayer capacity of a chromatographic packing along with a thermodynamic equilibrium constant for its adsorption [11] using the non-approximated expression of the electrostatically modified Langmuir adsorption isotherm and the rigorous surface potential of the non-approximated Gouy-Chapman (G-C) theory. IPC retention modeling can also reliably estimate ion-pairing equilibrium constants of analytes and IPRs [12].

IPC plays a fundamental role also in elucidating the thermodynamics of the pairing process. Experimental evidence to date supports a physico-chemical description of the hydrophobic ion-pairing process at variance with the electrostatic Bjerrum-type modelling of ion-pairing (see Sections 2.7.6 and 3.1.2).

IPC proved valuable for estimating peptide hydrophobicity [13]. Pharmaceutical science utilized IPC to monitor rat serum esterase activities [14] and also to analyze relationship between plasma concentrations at the end of infusion and toxicity profiles of fixed-dose-rate gemcitabine plus carboplatin [15]. An IPC trap was also used in an online desalting-mass spectrometry system. This system allows ionic compounds in a nonvolatile buffer to be introduced into a MS for strutural elucidation. The trap column was equilibrated with a volatile IPR, the target analyte and the nonvolatile buffer ions (phosphate and sodium ions) were transferred into the trap column, but only the target analyte that interacts with the volatile IPR can be retained ; phospahte buffer ion were eluted from the trap column and the target analyte was eluted by oragnic solvent in a backflush mode and introduced into the MS.

REFERENCES

1. Niederstatter, H., Oberacher, H., and Parson, W. Highly efficient semi-quantitative genotyping of single nucleotide polymorphisms in mitochondrial DNA mixtures by liquid chromatography electrospray ionization time-of-flight mass spectrometry. *Int. Congr. Ser.* 2006, 1288, 10–12.
2. Elhawary, N.A., Shawky, R.M., and Elsayed, N. High-precision DNA microsatellite genotyping in Duchesne muscular dystrophy families using ion-pair reversed-phase high performance liquid chromatography. *Clin. Biochem.* 2006, 39, 758–761.
3. Oberacher, H. et al. Direct molecular haplotyping of multiple polymorphisms within exon 4 of the human catechol-O-methyltransferase gene by liquid chromatography–electrospray ionization time-of-flight mass spectrometry. *Anal. Bioanal. Chem.* 2006, 386, 83–91.
4. Song, R. et al. Analysis of single nucleotide polymorphism sites in exon 4 of the p53 gene using high-performance liquid chromatography/electrospray ionization/tandem mass spectrometry. *Eur. J. Mass Spectrom.* 2006, 12, 205–211.
5. Lam, C.W. Analysis of polymerase chain reaction products by denaturing high-performance liquid chromatography. *Meth. Mol. Biol.* 2006, 336, 73–82.
6. Pavlova, A. et al. Detection of heterozygous large deletions in the antithrombin gene using multiplex polymerase chain reaction and denatured high performance liquid chromatography. *Haematologica* 2006, 91, 1264–1267.
7. Shi, R. et al. Temperature-mediated heteroduplex analysis for detection of drug-resistant gene mutations in clinical isolates of *Mycobacterium tuberculosis* by denaturing HPLC, SURVEYOR nuclease. *Microb. Infect.* 2006, 8, 128–135.
8. Udaka, T. et al. An Alu retrotransposition-mediated deletion of CHD7 in a patient with CHARGE syndrome. *Am. J. Med. Genet. A.* 2007, 143, 721–726.
9. Oberacher, H., Niederstätter, H., and Parson, W. Liquid chromatography–electrospray ionization mass spectrometry for simultaneous detection of mtDNA length and nucleotide polymorphisms. *Int. J. Legal Med.* 2007, 121, 57–67.
10. Gross, E. et al. Allelic loss analysis by denaturing high-performance liquid chromatography and electrospray ionization mass spectrometry. *Human Mutat.* 2007, 28, 303–311.
11. Cecchi, T. Use of lipophilic ion adsorption isotherms to determine the surface area and the monolayer capacity of a chromatographic packing, as well as the thermodynamic equilibrium constant for its adsorption. *J. Chromatogr. A.* 2005, 1072, 201–206.
12. Cecchi, T. Influence of chain length of the solute ion: a chromatographic method for the determination of ion-pairing constants. *J. Sep. Sci.* 2005, 28, 549–554.

13. Mant, C.T. and Hodges, R.S. Context-dependent effects on the hydrophilicity and hydrophobicity of side chains during reversed-phase high-performance liquid chromatography: implications for prediction of peptide retention behaviour. *J. Chromatogr. A.* 2006 1125, 211–219.
14. Koitka, M. et al. Determination of rat serum esterase activities by an HPLC method using S-acetylthiocholine iodide and p-nitrophenyl acetate. *Anal. Biochem.* 2008, 381, 113–122.
15. Wang, L.R. et al. The efficacy and relationship between peak concentration and toxicity profile of fixed-dose-rate gemcitabine plus carboplatin in patients with advanced non-small-cell lung cancer. *Cancer Chemother. Pharm.* 2007, 60, 211–218.
16. Yoshida, H. et al. On-line desalting mass spectrometry system for structural determination of hydrophilic metabolites, using a column switching technique and a volatile ion-pairing reagent. *J. Chromatogr. A.* 2006, 1119, 315–321.

17 Conclusions and Future Research Needs

At the dawn of the new century, analytical chemists faced great pressure to support massive advances in the food, life science, medicine, pharmacology, toxicology, environmental, agricultural, and biological sectors and also address fundamental security and safety issues. IPC proved to be a versatile addition to the toolboxes of separation scientist to handle the constantly increasing numbers and complexities of samples. In the following we expound potential future directions IPC, based on promising and exciting development.

The breakthrough of novel classes of IPRs (chaotropic additives and ILs) challenged the theoretical description of the dependence of retention on typical optimization parameters impose order on the complex welter of the theory is asked to retention patterns, and artificial neural networks are a versatile tool to describe them.

The potential of ILs as stationary phase modifiers must be investigated, based on the positive effects noted when ILs were added to mobile phases. Surface-confined IL stationary phases are intriguing because effective separations are achieved with aqueous mobile phases almost without organic solvents. Moreover in silica-based long-chain alkylimidazolium stationary phases, the alkyl chain length that does not impact electrostatic interactions may affect the hydrophobicity of the stationary phases. The modification of the reversed phase packing is similar to that obtained under dynamic adsorption of IPRs and the theoretical modelling of the interactions can take advantage of IPC theory.

We predict that the physico-chemical phenomenon known as ion pairing upon which IPC is based will be further exploited in different separative techniques such as CE and SFC because the modifications of analyte charge status and hydrophobicity are effective for achieving separation.

While separation media with zwitterionic functionalities for liquid chromatography have been investigated for many years, the use of zwitterionic IPRs for the separation of charged analytes was not explored but can be recommended. We submit that the influence of IPR concentration on the retention of neutral analytes (that was ignored for a long time because it is weak) may be beneficial when used for difficult-to-separate isomers and related substances. We also suggest the use of non-ionic additives to fine tune analyte retention. pH-tunable IPRs may be devised to switch from IPC to a reversed phase mechanism when a mixture of ionic, ionizable, and neutral solutes must be separated. Temperature is still an underutilized optimization parameter and it can be easily predicted that it will be expedient for achieving ultra fast separation. We believe and hope that the efforts of the community of separation scientists will further advance this exciting separation strategy and related techniques.

Index

Printed and bound by CPI Group (UK) Ltd, Croydon, CR0 4YY

18/10/2024

01776245-0007